Musical Vitalities

NEW MATERIAL HISTORIES OF MUSIC

A series edited by
James Q. Davies and Nicholas Mathew

Musical Vitalities

Ventures in a Biotic Aesthetics of Music

HOLLY WATKINS

The University of Chicago Press
Chicago and London

The University of Chicago Press, Chicago 60637
The University of Chicago Press, Ltd., London
© 2018 by The University of Chicago
All rights reserved. No part of this book may be used or reproduced in any manner whatsoever without written permission, except in the case of brief quotations in critical articles and reviews. For more information, contact the University of Chicago Press, 1427 E. 60th St., Chicago, IL 60637.
Published 2018
Printed in the United States of America

27 26 25 24 23 22 21 20 19 18   1 2 3 4 5

ISBN-13: 978-0-226-59470-5 (cloth)
ISBN-13: 978-0-226-59484-2 (e-book)
DOI: https://doi.org/10.7208/chicago/9780226594842.001.0001

Library of Congress Cataloging-in-Publication Data

Names: Watkins, Holly, 1972– author.
Title: Musical vitalities : ventures in a biotic aesthetics of music / Holly Watkins.
Description: Chicago ; London : The University of Chicago Press, 2018. | Includes bibliographical references and index.
Identifiers: LCCN 2018017934 | ISBN 9780226594705 (cloth : alk. paper) | ISBN 9780226594842 (e-book)
Subjects: LCSH: Music—Philosophy and aesthetics. | Nature in music.
Classification: LCC ML3845 .W27 2018 | DDC 781.1/7—dc23
LC record available at https://lccn.loc.gov/2018017934

♾ This paper meets the requirements of ANSI/NISO Z39.48-1992 (Permanence of Paper).

Contents

	Introduction	1
1	Reanimating Musical Organicism	15
2	Formalism's Flower	41
3	Schopenhauer's Musical Ecology	66
4	The Floral Poetics of Schumann's *Blumenstück*, op. 19	85
5	Music between Reaction and Response	112
6	On Not Letting Sounds Be Themselves	132

Acknowledgments 155
Notes 157
Bibliography 181
Index 197

Introduction

Music is the art of enlivening sounds. Music transforms the acoustic results of material excitation—of keys striking hammers, bows traveling across strings, columns of air vibrating in larynxes—into signs of excitation unbounded, depersonalized, writ large. Musical sounds constitute a lexicon of arousal whose interpretive possibilities span physical, physiological, emotional, and conceptual realms. Yet music also teaches us that these realms cannot be separated so easily. Music's play of sonic and formal energies may simulate the release of mechanical or inorganic energies, but historically music has been a partner to expressly organic pursuits. The danceable tune, the singing line, the invigorating rhythm, the seductive timbre—music not only enhances feelings of vitality but also projects its own sense of animation.[1] Even music that soothes and relaxes does so by entraining us to alternative metabolisms sonically realized. "We hear music as a manifestation of vitality," writes philosopher Kathleen Marie Higgins, "and part of our enjoyment is empathy with its liveliness."[2] Empathizing with music, we encounter a form of animation—not biological, but not merely illusory either—that thrives where human, organic, and inorganic energies cross over and shade into one another.

Whose life resonates in music? The answer appears to be obvious: the lives of those who create it, play it, listen to it, dance to it, daydream to it. But music's liveliness is not reducible to that of the agents responsible for its sounding or its reception. Music is emergent, so that when, say, a melody is performed adequately, it becomes something more than just a sequence of sounds—it becomes something holistic rather than additive. That which coalesces out of tones or beats, such as a metrical pattern or recurring refrain, appears to take on a lifelike, self-maintaining character. Listeners are necessary participants in this phenomenon, but they are not its sole point of origin. What music does and what listeners hear are mutually constitutive.

Emergence, however, does not fully account for music's liveliness. The life of music—tenuous, metaphorical, contingent, and mortal—has multiple sources: not only the self-sustaining character of musical patterns but also the phylogenetic significance of hearing and the cross-modal interpretation of musical actions. Sounds alert us to dynamic forces in the environment, to the presence of predators and prey, to dangers and lures both animate and inanimate. Sounds are ambiguous: regular patterns (*drip-drip-drip* or *inhale-exhale*) and singular events (a *boom* or a *scream*) result from both unintended physical events and the intentional actions of living beings. Music retains that ambiguity; music is the art of possibly animate things. Music spurs us to imagine creating, being, or undergoing an almost endless variety of dynamic movements that, especially in the case of instrumental music, need not be heard as expressions of human subjectivity or embodiment.[3] Such imaginative work generally does not take place on a conscious level. Listening to music, we unconsciously experiment with being other. Music creates a multitude of virtual worlds, or virtual configurations of space and time, that listeners can vicariously experience as alternative forms of embodiment, affect, spirit, thought, or some combination thereof.[4] Music makes us feel more present and embodied, but it also carries us away. In either case, music affords experiences of selfhood that are broadly distributed across the terrain of body and mind. Music both diversifies the self and extends it toward other selves in motion, whether real or imaginary, human or not.

This book is about the vitality of music in all the enigmatic senses I have just outlined, which cluster around two of music's most cherished aptitudes: to stimulate and simulate life. Roughly equal parts philosophy of music, history of aesthetics, and analytically informed criticism, *Musical Vitalities* engages with a body of thought spanning the humanities and the sciences in search of a more expansive understanding of musical form and meaning.[5] The book's interlocutors include the theory and philosophy of embodied cognition (Francisco Varela, Evan Thompson); systems theory (Gregory Bateson, Niklas Luhmann); biological and anthropological semiotics (Terrence Deacon, Eduardo Kohn); critical animal and plant studies (Donna Haraway, Michael Marder); and contemporary philosophy and aesthetics (Elizabeth Grosz, Gilles Deleuze and Félix Guattari).[6] *Musical Vitalities* also crosses paths with familiar but neglected contributions to aesthetics such as Susanne Langer's *Feeling and Form*, which considers music a peculiarly efficacious means of symbolizing vital existence. Whereas Langer's theory remains largely idealist and humanist in spirit, however, my study seeks to elucidate

INTRODUCTION

how music brings its human practitioners into real and imagined contact with more-than-human vitalities.[7] By the latter, I mean aspects of existence that are experienced by humans but not limited to them, such as being subject to vibration and other physical forces, periodic motion (the breath, the heartbeat, the gait) as well as entrainment to external periodicities (the cycles of days, months, and years), and semiotic practices whose formal properties span multiple arenas of living expression. We most easily imagine sharing these experiences with our closest animal relatives, but the reach of the dynamic principles at work extends much further—to the plant kingdom and even the realm of inanimate objects.[8]

Although it has become virtually obligatory for musicologists to refrain from discussing culturally specific musical practices in terms of their naturalness or relation to the natural world, such restraint—useful though it has been in checking the universalist impulses of Western thought—has allowed a set of biases to take hold that seem to me increasingly untenable.[9] By tacitly endorsing human exceptionalism and its distinctions between human culture and the rest of what goes on in nature, the bulk of musicology today chugs along in an almost exclusively humanistic mode, ignoring developments in sister disciplines such as music cognition and biomusicology as well as more distantly related fields like ethology and biosemiotics.[10] As a result, music's structural resemblances to the songs of nonhuman species, the intricacies of music's physiological impact on the animal bodies of its listeners, and the many analogues between music's formal processes and those unfolding elsewhere in the natural world rarely come under consideration outside specialist venues devoted to animal behavior, music perception, and ecomusicology.[11] The goal of this book is neither to canvass on behalf of a particular subfield nor, conversely, to argue for the absorption of musicology into the more diffuse category of sound studies. Instead, *Musical Vitalities* seeks to counteract both the fragmentation of intellectual inquiry into isolated niches and the jettisoning of musicological tools capable of illuminating connections between human music making and the dynamic expressiveness, or expressive dynamism, of the world at large. It is precisely by preserving some of musicology's most distinctive features—namely, the formal terminology it uses to describe patterned sound and a vocabulary geared to nuances of affect and meaning—that this book finds the leverage to trouble persistent divisions between the humanities and the sciences. Add to that my occasional recourse to "gnostic" interpretation, my interest in treating human musicking as a subset of the cultural activities and "biotic arts" of other animals, and my sheer intellectual greediness, and the mixture of disciplinary voices *Musical Vitalities* rouses into conversation is nearly complete.[12]

The biotic aesthetics that flourishes in this polyvocal critical space makes possible the reappraisal of a body of literature that *did* make ample use of naturalistic imagery, often to the chagrin of later interpreters keen to expose its prejudices. Through close readings of (mostly) nineteenth-century German music and aesthetic writings that rely on a range of analogies between music and nature, *Musical Vitalities* seeks to both rekindle the intellectual potential of a literature often decried as exhausted of contemporary significance and rejoin the humans at the center of the humanities with the nonhumans who feature so frequently in their imaginations. The book's six chapters comprise a series of interpretive alloys that fuse historical discourses such as organicism, formalism, and Schopenhauer's aesthetics with insights drawn from systems theory, semiotics, and the life sciences. My points of departure are allusions to music as an art whose formal processes resemble the growth of plants, whose beauty is intimately related to natural beauty, whose sonic fluctuations open out onto larger movements of energy and will, whose vibratory and affective enchantments appeal to regions of the human body that elude conscious surveillance and control, and whose strategies of signification have much (but not everything) in common with modes of animal communication. Several of the chapters pair these allusions with music—most of it by Robert Schumann—that takes some aspect of nonhuman nature as its thematic conceit. Why Schumann? Perhaps because his music never quite lived up to the standard of humanist self-overcoming epitomized by Beethoven's heroic works.[13] And perhaps because, as I discovered in previous research, there is a certain semiotic mutability about Schumann's music that makes it amenable to conceptual reinvention.[14] The composer's oeuvre accordingly presents numerous opportunities to ponder less anthropocentric ways of construing musical vitality.

The primary concern of *Musical Vitalities* is not with what Schumann's or other composers' musical evocations of the natural world tell us about how nature was conceived in nineteenth-century Germany.[15] Although the book shares an interest in points of contact between music and the natural world with the field of ecomusicology, it does not, in contrast to much work in that field, focus its attention on depictions of or interactions with any specific environments.[16] Rather, I place the music and thought of earlier eras on a collision course with more recent philosophical and scientific thought to develop a musically inflected notion of the human that embraces the evolutionary heritage we share with other creatures and our entangled planetary fates. To this end, *Musical Vitalities* locates chinks in the Germanic tradition of musical aesthetics through which less asymmetrical relationships between humans and nonhumans, and less dualistic modes of thinking about cultural artifacts and natural entities, can be glimpsed.

This project arose in sympathy with multidisciplinary efforts to bring humanistic inquiry in line with ways of thinking about human existence currently being explored by scholars affiliated, explicitly or not, with posthumanism.[17] Since *Musical Vitalities* continues to make use of recognizably humanistic interpretive methods, and since I am disconcerted by the increasingly cozy relationship between academic fashion and the neoliberal directive of constant innovation, I refrain from hoisting the banner of this or any other scholarly movement over the book's intellectual terrain. In any case, the book's preoccupation with historical subject matter represents a departure from how music is usually treated by posthumanist writers, who refer mainly to modern composers (John Cage is a special favorite) and popular repertories, as if only contemporary music could bear the application of contemporary thought.[18] I also differ from music scholars for whom posthumanism and new materialism primarily afford means of illuminating musical culture's technological conditions of possibility.[19] While I do not deny the importance of those conditions, nor do I wish to draw a firm boundary between human bodies and the tools they shape and by which they are shaped, my study nonetheless remains focused on instances where the self-manifesting movement of living things (*physis*) is held up as the ideal to which human making (techne or *poiēsis*) aspires.[20]

Despite these caveats, several of the book's enabling presuppositions are clearly related to those of posthumanism, especially its more ecologically oriented varieties. For example, I am curious how a more realistic picture of the interdependence of conscious thought, unconscious cognition, and what N. Katherine Hayles calls the cognitive nonconscious might alter our assessment of human endeavors, and of music in particular. Humanism, with its emphasis on the empowering faculty of reason, is preoccupied with the fruits of human consciousness, using these as the basis for an exceptionalism that places all other beings below us on the scale of value and ethical consideration. When humanists limit themselves to registers of meaning accessible solely to human consciousness, they sever their objects of inquiry from the larger physiological and ecological settings from which they emerge. Rejecting the notion that expanding the purview of the humanities beyond conscious activity threatens the possibility of critique, Hayles writes, "Consciousness alone cannot explain why scholars choose certain objects for their critique and not others, nor can it fully address the embodied and embedded resources that humanities scholars bring to bear in their rhetorical, analytical, political, and cultural analyses of contemporary issues. Without necessarily realizing it, humanities scholars have always drawn upon the full resources of human cognitive ecologies, both within themselves and within their interlocutors."[21] Diminishing the importance granted to conscious thought in conceptions of the human not only sets the embodied nature of critical endeavors in clearer relief but also helps us conceptualize aesthetic experience itself as something much more complicated than mind-to-mind transactions between creators and receivers.[22]

Posthumanism further diversifies and complicates the human by taking into account our physiological and cultural commonalities with other creatures, the relational networks between species and environments on which our survival depends, and our own status as multispecies consortia made up of human cells and at least an equal number of bacterial cells, not to mention parasites and other hangers-on.[23] The scope of the humanities stands to be dramatically augmented once we truly confront the fact that humans do not exist in a vacuum, that humans depend on a vast ecological web of relations for their survival, that the human does not consist solely in what we think we are, and that the human genome is 99 percent similar to that of our closest primate relatives, chimpanzees and bonobos (not to mention 60 percent like that of fruit flies).[24] It is as if humanists were stuck with the unflattering slogan "we are the 1 percent." This is not to deny that there are differences between humans and other species—after all, about thirty-five million genetic variances occupy that 1 percent of difference—but to advocate for a strategic redistribution of the vocabulary of sameness and difference.[25] Pondering the similarities between humans and other creatures opens the door to greater interest in and care for nonhuman others; in addition, we could benefit from extending the discussion of difference and diversity beyond their customary humanistic confines. Human cultural and biological differences, despite being sources of perpetual political conflict, are far exceeded by the staggering biological diversity of earthly life. Although fighting social injustice often depends on increasing rather than decreasing sensitivity to human difference, it is probably also the case that a concept of humanity that is both more inclusive and less differentiated from the many other species on whom human flourishing depends will be essential in combating the potentially catastrophic effects of anthropogenic global warming.[26] As only one of about fifty-five hundred mammal species, a million animal species, and at least two million species all told—and these numbers represent only documented lifeforms—humans stand to learn a lot about diversity from the natural world. The mind-boggling variety of gender-specific behaviors and means of sexual reproduction, for instance, easily demolishes human notions of what is "natural" in these arenas.²⁷ At the very least, the interests of humans provide reason enough to value planetary biodiversity, since it is crucial to our long-term survival as well as a key source of existential and aesthetic wonder.²⁸

To be sure, such locutions as "the natural world," as used earlier, can have troubling implications when juxtaposed with specifically human activities. Although I occasionally find it necessary in this book to distinguish human pursuits from those found elsewhere in the world, such rhetorical gambits should not be taken as suggesting that humans somehow exist apart from nature.²⁹ "There is but one life," writes philosopher Catherine Malabou, "one life only."³⁰ Malabou urges both humanists and scientists to explore the continuity between biological and symbolic manifestations of sentience, a project whose groundwork has been laid by Terrence Deacon in his extraordinary book *Incomplete Nature: How Mind Emerged from Matter*. The perspectives of these authors accord with the more general posthumanist suspicion of the boundaries Western thought has drawn between humanity and nature. For Bruno Latour, an imaginary but conceptually potent "Modern Constitution" upholds categorical distinctions between the social world of humans and the natural world of nonhumans.³¹ Constructivist views of human social life that place such complex phenomena as sexuality and gender identity entirely under the control of culturally determined modes of thought only exacerbate the sense of division between humans and nature.³² The social activities of humans, who are themselves products of natural selection, cannot be simply cordoned off from the rest of the natural world by fiat. This humanist maneuver is not only belied by the cultural transmission of behaviors among animals but also largely the conceptual product of the modern West and therefore subject to contention.³³ In addition to Latour's interrogation of Western scientific practices, ethnographic studies of non-Western societies offer a powerful means of challenging assumptions regarding the opposition between nature and culture; ethological studies (especially those of primates) are another, as are studies of the impact that microbiotic communities residing in, on, and around human bodies have on socially pertinent phenomena such as mood regulation and the onset of disease.³⁴ We have seen nature, and it is us.

While it can be useful to distinguish between the natural and the human-made, real conundrums arise when one tries to determine where the creativity of nature (*physis*) ends and the creativity of humans (techne or *poiēsis*) begins. Humans may intervene in *physis*, as when they selectively breed or genetically modify plants and animals, but they are not themselves the cause of growth. Yet since humans too are a product of *physis*, and, as Aristotle maintained, since techne is a naturally occurring aptitude and therefore also related to *physis*, one cannot enforce absolute distinctions between the two kinds of creating, nor between, in James Currie's terms, artifice and the real.[35] Although humans can identify instances of *physis* in which they have taken no part, it is much harder to eliminate *physis* altogether from techne.[36] This is arguably what eighteenth- and nineteenth-century commentators were getting at when they likened the obscure workings of human creativity to unconscious, plant-like growth.[37]

If the ecological crises unfolding across the globe have lent new urgency to the project of rethinking the human, the conceptual resources for tackling this project have been around, in some form, for many decades, even centuries. Like any "-ism," posthumanism is a recrudescence of attitudes with many antecedents. The nineteenth-century philosophers Arthur Schopenhauer and Friedrich Nietzsche, for example, realized that the musings of the conscious "I" represent only a small portion of human existence, and both of them left ample room in their writings for pondering the animal- and plant-like aspects of human physiology. I will return to Schopenhauer in chapter 3; for now, I would like to spend a few moments with Nietzsche to show that radically rethinking the human does not require the imprimatur of a current scholarly trend. "The greater part of conscious thinking," Nietzsche remarks in *Beyond Good and Evil*, "must be counted amongst the instinctive functions," a statement that immediately casts doubt on conventional humanist assumptions.[38] Nietzsche continues:

> The people on their part may think that cognition is knowing all about things, but the philosopher must say to himself: "When I analyze the process that is expressed in the sentence 'I think,' I find a whole series of daring assertions, the argumentative proof of which would be difficult, perhaps impossible: for instance, that it is *I* who think, that there must necessarily be something that thinks, that thinking is an activity and operation on the part of a being who is thought of as a cause, that there is an 'ego,' and finally, that it is already determined what is to be designated by thinking—that I *know* what thinking is.... With regard to the superstitions of logicians, I shall never tire of emphasizing a small, terse fact, which is unwillingly recognized by these credulous minds—namely, that a thought comes when 'it' wishes, and not when 'I' wish; so that it is a *perversion* of the facts of the case to say that the subject 'I' is the condition of the predicate 'think.'"[39]

What becomes of human self-esteem when thinking or, more pointedly, rationality is viewed as inseparable from a greater psychological and physiological multiplicity that includes, for Nietzsche, "willing" and "feeling," not to mention more mundane processes such as digestion and elimination? It is not that self-esteem evaporates altogether—Nietzsche's own glowing self-evaluation in *Ecce Homo* is proof enough of that—but that (modern, Western) humanity's self-image must be recalibrated to account for the vagaries of an organic existence profoundly affected by nutrition, climate, physical activity, and state of health, among other factors. Mental life does not simply go about its business irrespective of physiological and environmental concerns, as Nietzsche bluntly asserts in *The Genealogy of Morals*:

> When any one fails to get rid of his "pain in the soul," the cause is, speaking crudely, to be found *not* in his "soul" but more probably in his stomach . . . A strong and well-constituted man digests his experiences (deeds and misdeeds all included) just as he digests his meats, even when he has some tough morsels to swallow. If he fails to "relieve himself" of an experience, this kind of indigestion is quite as much physiological as the other indigestion—and indeed, in more ways than one, simply one of the results of the other.[40]

However hyperbolic one may find Nietzsche in his proto-Freudian moods, he does at least illustrate, with the help of a little dark humor, that one's thoughts are related to more than just one another. To put the matter in systems-theoretical terms, the psychic and digestive systems are coupled in such a way that each may perturb, though not fully determine, the other.

Elsewhere in *Beyond Good and Evil*, Nietzsche shows how the attempt to distinguish humans from nature lands one in a thicket of contradictions. He criticizes the Stoic desire to live "according to Nature" as both impossible and inevitable, a paradox that arises when the concept of nature fails to include nature's own internal antagonisms. "Imagine to yourselves," Nietzsche begins, "a being like Nature, boundlessly extravagant, boundlessly indifferent, without purpose or consideration, without pity or justice, at once fruitful and barren and uncertain—how *could* you live in accordance with such indifference?" And yet, living beings, by their very nature, struggle against these overarching tendencies: "To live—is not that just endeavoring to be otherwise than this Nature? Is not living valuing, preferring, being unjust, being limited, endeavoring to be different?" Living, in short, depends on the assertion of difference in the face of indifference, and this assertion is itself part of nature. Nietzsche continues, "Granted that your imperative, 'living according to nature,' means actually the same as 'living according to life'—how could you do *differently*? Why should you make a principle out of what you yourselves are, and must be?"[41] The upshot is that appeals to nature or the natural are not fine-grained enough to provide real ethical guidance as to how humans ought to live. We are part of nature, yet we find ourselves both sustained and threatened by it, drawn toward and repelled by it, praising its benevolence and lamenting its heartlessness. It takes only one bacterial infection or residential infestation to strain human feelings of oneness with nature.

Nature, in short, is the arena of contest between irreconcilable values as well as cooperation toward common ends. Lawrence J. Hatab argues that Nietzsche's emphasis on values offers a refreshing alternative to scientific naturalism, which drains the natural world of meaning and value in its quest for the empirical verification and mathematical formalization of physical laws.[42] By contrast, all forms of life, in Nietzsche's view, embody a perspective on the world, and perspective implies value: what is good for the tick is not good for the animal on whom it feeds. As this by now familiar example suggests, Nietzsche's thought anticipates that of early twentieth-century biologist Jakob von Uexküll, who posited that every living creature constructs an environment (or *Umwelt*) using its own peculiar sensory and cognitive apparatus.[43] The process of construction in turn invests the world with value and meaning, and this valuation holds for all organisms, not just humans. While human language—and with it the ability to create elaborate networks of concepts, refer to things not present, and invent fictional or hypothetical realities—represents an unusual extension of the power of construction via symbolic thought, it is only one manifestation of the way all life forms construct their worlds.

Importantly, however, constructivism need not entail the belief that there is no real world—only that an untotalizable reality is selectively perceived by any species or organism.[44] In other words, no creature constructs its world in a vacuum. Although the nature that each creature experiences is different, it is still legitimate, I believe, to refer to nature as just this untotalizable reality. In this sense, I find myself allied more with what Kate Soper refers to as "nature-endorsing" thinkers rather than "nature-skeptical" ones.[45] Nietzsche, for his part, shows the absurdity to which constructivism, as a kind of extreme Kantianism, invariably leads. He maintains that "the sense-organs are *not* phenomena in the sense of idealistic philosophy," meaning that they do not belong to the realm of mere appearance. In response to those who hold "even that the external world is the work of our organs," Nietzsche replies, "But then our body, as part of this external world, would be the work of our organs! But then our organs themselves would be the work of our organs!" The only way out of this reductio ad absurdum is to conclude that "the external world is *not* the work of our organs"—a statement Nietzsche concludes with a question mark, as if to consign the whole matter to the rarefied air of idle speculation.[46]

In an introduction to a recent collection of essays, Vanessa Lemm sums up the relevance of Nietzsche's writings for contemporary ecological and posthumanist thought: "The continuity between human life and the life of all organic and inorganic matter unsettles our anthropocentric conception of the world and shows that human culture and civilization must be understood as part and parcel of the greater order of the totality of life."[47] Viewing human culture as something like what Nietzsche called "transfigured *physis*"—a phrase that renders techne continuous with *physis* rather than separate from it—necessitates contemplating the distinctiveness of human agency against a greater background of interspecies similarity.[48] *Musical Vitalities* furthers this endeavor by exploring how analogies between musical and natural processes, which appear repeatedly in the literature on musical aesthetics, encourage us to develop modes of thinking that challenge presumed divisions between cultural artifacts and natural entities. Each chapter is a variation on this theme, and each experiments with imbuing analogies between music and nature with contemporary ecological and critical significance.

Chapter 1, "Reanimating Musical Organicism," revisits the legacy of organicism to discover fresh critical potential in a discourse currently maligned as a relic of Austro-German chauvinism. Even a casual perusal of the primary literature shows that organicism was always beset by internal tensions and unresolved issues, many of which stem from the peculiarities of the organisms that typically served as this literature's inspiration: plants. Touching on figures ranging from Immanuel Kant, Johann Wolfgang von Goethe, and Theodor Adorno to the music critics Christian Friedrich Michaelis, E. T. A. Hoffmann, and Eduard Hanslick, I show that organicist writings, many of which compared the real-time unfolding of musical works such as Beethoven's Fifth Symphony and Wagner's *Tristan und Isolde* to plant growth, continue to raise questions about the lives and identities of both organisms and artifacts as well as the relationships between these different expressions of vitality. Drawing on the social systems theory of Niklas Luhmann, whose conceptual and analytical tools deftly mediate between organic and cultural modes of organization, I offer a series of novel perspectives on the quasi-organic traits of musical form and stylistic change, which I then use to sow the seeds of a new organicism that embraces the organisms at its heart.

Chapter 2, "Formalism's Flower," elaborates on the theme of form as it appears in two key contributions to the philosophy of natural and musical beauty: Kant's *Critique of Judgment* and Hanslick's *On the Musically Beautiful*. In keeping with the current interest among environmental philosophers in remedying the nearly exclusive focus of post-Kantian aesthetics on the human arts, the chapter examines how Kant's and Hanslick's reflections on beauty highlight the formal and experiential ground shared by music and nature, ground that neither thinker explored in detail (Hanslick, in keeping with his idealist orientation, spent far more time describing what differentiates the two). Drawing on ideas advanced in both treatises, I seek to articulate a formalism that neither locates the value of musical works exclusively in relationships between notes nor excludes historical, contextual, or personal factors from consideration. Instead, I develop conceptual strategies in which aspects of form and beauty serve to illustrate processual and dynamic features of both music and nature, strategies I employ in a discussion of arabesque and its musical analogues. As a decorative art in which the mimesis of vegetal forms and energies is crossbred with human geometrical precision, arabesque points the way toward a naturalistic music criticism that nonetheless remains focused on the peculiarities of its chosen art, as I show in an analysis of Robert Schumann's *Arabeske*, op. 18 (1839) for solo piano.

Chapter 3, "Schopenhauer's Musical Ecology," takes its inspiration from the philosopher's famous comparison of the registers of polyphonic musical textures (soprano, alto, tenor, and bass) to the "grades" of earthly existence. Moving from the high notes of the melody down to the deep tones of the bass was comparable, Schopenhauer thought, to traversing human, animal, vegetal, and inorganic domains. Despite his reputation as a metaphysician, Schopenhauer helps us think across and beyond conventional distinctions between humans and nonhumans (including, in the context of this chapter, nonliving matter). Schopenhauer's recognition of the relatedness of all beings and the presence of mineral, vegetal, and animal grades of will in human bodies makes his philosophy well worth revisiting at a time of burgeoning interest in vital materialism and the "nonhuman turn."[49] By reading Schopenhauer's aesthetics against the metaphysical grain, so to speak, I demonstrate that his remarks on music clearly delineate the art's physical impact, even as he locates musical expression in a region beyond that of any particular body. Furthermore, I show that combining Schopenhauer's ecological conception of music with his multilayered notion of the body generates a surprisingly pluralistic account of musical experience, one whose scope includes material, organic, and psychic facets of existence. The chapter closes by reflecting on the by turns promising and problematic nature of Schopenhauerian transcendence in an era marked by global warming.

Chapter 4, "The Floral Poetics of Schumann's *Blumenstück*, op. 19," follows a somewhat different path than the other chapters of the book. Instead of staging an encounter between, say, complementary strains of Romantic and posthumanist thought, the chapter focuses on how flowers, those supreme representatives of nonhuman beauty, were woven into nineteenth-century conceptualizations of gender, art, and transcendence. Thanks in part

INTRODUCTION

to Schumann's own disparaging remarks, *Blumenstück* (1839), a short piano piece similar to the *Arabeske*, has been viewed as a fairly straightforward effort to appeal to amateur consumers—especially female consumers—of domestic piano music. The piece's mixed aesthetic status is closely linked to the similarly ambivalent standing of flowers (and the genre of flower painting to which Schumann's title alludes) in early nineteenth-century Germany. Yet although flowers were normally thought to be emblematic of women and the conventionalized expression of sentiment, they also constituted a remarkably evocative symbol in Romantic literature. Sentimental (or *Biedermeier*) and Romantic discourses of the flower converged in the trope of *Blumensprache* (the language of flowers), a signifying practice developed in popular manuals cataloging the meanings of flowers and referenced in the more esoteric settings of Schumann's criticism, Hoffmann's tales, and Heinrich Heine's poetry. In each of these venues, flowers served as nonhuman conduits for imaginary travel between mundane and transcendent realms. Drawing on the work of Friedrich Kittler, I elaborate on related dualities in Schumann's *Blumenstück*. With its conflicting imperatives of pleasure and instruction, congenial melody and motivic intertwining, the piece conflates aesthetic categories in a manner that undermines traditional notions of both organicism and generic classification.

Chapter 5, "Music between Reaction and Response," evaluates music's capacity to thwart conceptions of the human based on the sovereign power of rationality from a range of philosophical, critical, and scientific standpoints. Music's problematic blurring of the boundaries separating human from nonhuman bodies has long been recognized, as two Greek myths attest: Orpheus made music that inspired humanlike attention in animals, trees, and stones, while the Sirens reduced passing sailors to the level of animals incapable of resisting their song. Recast in terms employed by Jacques Lacan and criticized by Jacques Derrida, these myths portray music as calling forth a response in creatures thought merely able to react and, contrariwise, stripping away the capacity for response in humans, leaving nothing but reaction in its place. While music often provokes highly refined cognitive and emotional responses, it also acts upon the body in a variety of ways, many of them involuntary—a fact that has struck music's advocates as alternately promising and disturbing. After briefly considering the debate between Lacan and Derrida, I revisit eighteenth- and nineteenth-century commentaries by the philosophers and critics Johann Georg Sulzer, Johann Gottfried Herder, and Hanslick so as to illuminate persistent anxieties over the admixture of reaction and response in musical listening—an admixture that carries with it the further threat of confusion between animal and human modes of experience. Turning to more recent studies in music perception and ethology, the chapter weaves research on the physiological reactions involved in musical responsiveness into a philosophical perspective on the expressiveness of sound that accommodates the communicative arts of both humans and animals.

Chapter 6, "On Not Letting Sounds Be Themselves," begins with a critique of the familiar modernist notion of "sounds themselves," which crops up in writings by composers ranging from John Cage to Pierre Schaeffer to John Luther Adams. Cage's 1957 essay "Experimental Music," for instance, famously enjoined composers to "set about discovering means to let sounds be themselves" rather than continuing to use sound as a means of all-too-human expression.[50] On the face of it, lending an ear to sounds themselves seems to foster a more inclusive approach to sonic experience by refusing to honor putative distinctions between human-made sounds (including music) and sounds originating from nonhumans, living or otherwise. Yet this apparent catholicity arises from a rather strange understanding of sound in nature, one in which, as John Luther Adams puts it, natural sounds are considered "direct, immediate and non-referential."[51] Rather than representing an escape from signification, however, the natural world is positively saturated with signs. Expanding on applications of Peircean semiotics by Naomi Cumming, Gary Tomlinson, and Eduardo Kohn, this chapter turns a biosemiotic lens on the multilayered semiotics of music as manifested in Schumann's "Vogel als Prophet," a movement from the piano cycle *Waldszenen*, op. 82 (1849), and in compositions by Adams and the Norwegian composer Jana Winderen.

The abundant natural imagery running through nineteenth-century European music and aesthetic discourse testifies to the considerable impact that the vitality of nonhuman others has had on products of human imagination. There may be plenty we wish to reject in the writings of Hanslick, Schopenhauer, and others, and we may be convinced that it is no longer relevant to compose music like Schumann's. Yet ideas are renewable resources, and changing circumstances breathe new life into them. What follows might be described as the philosophy of a scavenger who scrounges around the scattered remains of nineteenth-century culture for life-sustaining tidbits, or perhaps as that of an aesthetic mutualist, whose embodied (meaning physical, affective, and cognitive) experiences of music and the natural world perpetually inform, enrich, and challenge one another. In any case, it is my hope that the chapters to come create opportunities to marvel anew at music's power to evoke more-than-human modes of embodiment as well as stimulate, transform, and complicate the vitalities of those humans who fall within its compass.

1

Reanimating Musical Organicism

Perhaps Theodor Adorno had the opening of *Tristan und Isolde*'s second act in mind when he upheld Richard Wagner's opera as an exemplary instance of musical organicism (example 1.1). Melodic lines in the winds twist this way and that like so many fronds and tendrils, proliferating across the introduction with all the tenacity of a weed.[1] Elaborating on the botanical metaphor in the 1961 essay "Vers une musique informelle" ("Toward an Informal Music"), Adorno remarked, "The minimal, as it were effortless, transition of semitone steps is regularly associated with the idea of growing plants, since it appears not to have been manufactured, but seems as if it were growing towards its final purpose without the intervention of the subject." In Adorno's reading, organicism originates in the impression that music is self-generating, that it is invested with an entelechy not unlike that of a living being. Chromaticism, by enhancing the directedness of musical motion, only increases what Adorno referred to as "the semblance of the organic as mediated by this language"— namely, the language of tonal music.[2]

Yet music need not be highly chromatic to inspire thoughts of organic life. The intricate motions of counterpoint put E. T. A. Hoffmann in mind of the "intertwining of mosses, weeds, and flowers," while the musical enthusiasts depicted in his *Kreisleriana* hear the voices of "trees, flowers, animals, stones, water" resounding in their favorite art.[3] Isolde too is drawn in by music's capacity to evoke not only organic but also inorganic nature. A few moments after the aforementioned passage, she and Brangäne argue over whether King Marke's hunting party has retreated safely into the night—the sign that Tristan can make his approach. Brangäne distinctly hears the men's horns, while Isolde hears only the sounds of the garden surrounding the two

EXAMPLE 1.1. Richard Wagner, *Tristan und Isolde*, act 2 prelude, mm. 29–55

women—wind in the leaves, water in the fountain. The orchestra's music tracks the difference between the two women's perceptions (example 1.2).

Unable to hear what her mistress hears, Brangäne complains, "You are deluded by the wildness of your desire into hearing only what you choose to."[4] In the wake of musicology's attempt to purge its lingering Germanocentrism, one may be tempted to level the same accusation at proponents of organicism, who maintain that the best music displays the integration of parts into a greater whole and develops in a manner similar to the growth of plants. As Lotte Thaler and Lothar Schmidt have recounted in studies of organicist discourse, critics and analysts such as Adolf Bernhard Marx, Eduard Hanslick, and Heinrich Schenker advanced authoritative but unstable claims

EXAMPLE 1.1. (*continued*)

regarding what makes music organic in hopes of establishing the superiority of formal principles exemplified by late eighteenth- and nineteenth-century Austro-German music, most of it (pace Wagner) instrumental.[5] Statements regarding the development of musical material out of a single seed and the reciprocity of parts and whole accrued rhetorical force despite the lack of consensus regarding their analytical demonstration.

Even worse, organicism today is commonly understood to entail distinctly regressive social and political values, thanks to its association with German nationalism and idealist theories of the state.[6] In his provocative book *The*

EXAMPLE 1.2. Wagner, *Tristan und Isolde*, act 2, scene 1, mm. 72–92

Ecological Thought, Timothy Morton writes that organicism does the "heavy lifting for homophobic Nature," meaning that its apparent emphasis on autonomy, boundedness, and internal versus external determination recapitulates values affiliated with a corrosive brand of heterosexual masculinity.[7] Yet if what is thought to constitute musical organicism varies with contemporary notions concerning organisms in general, as Thaler concludes at the end of her study, then perhaps it is time to take a different view of organicism's shortcomings.[8] What if the problem is not with the thesis that certain musical processes create a semblance of the organic but with the models of the organism brought in to give content to that semblance?

Nineteenth- and twentieth-century organicists favored analogies between

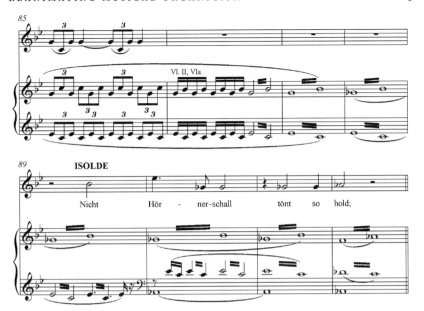

EXAMPLE 1.2. (*continued*)

music and either human beings or plants. How might the significance of such analogies change in the face of biological research on plant cognition (summarized by Michael Pollan in a recent article for the *New Yorker*) and philosophical attributions of mind to even the simplest organic beings?[9] Or in the wake of recognizing that humans share many biological features with both plants and animals? Philosopher Michael Marder alleges that "the human body and subjectivity alike are not pure expressions of Spirit but strange archives, surfaces of inscription for the vestiges of the inorganic world, of plant growth, and of animality—all of which survive and lead a clandestine afterlife in us, as us."[10] Thinking the human through the plant and the plant through the human clears the ground for a species of organicism that dispenses with humanism's androcentric conceits and prompts renewed reflection on the points of connection between musical and organic processes.

After critiquing the conventional appeals to wholeness and end-oriented development usually associated with organicism, this chapter reconsiders some of the formal features that have motivated organicist analysis from the standpoint of biological and social systems theory. For the German sociologist Niklas Luhmann, social systems, like biological systems, are self-organizing. Social systems reproduce themselves by way of recursive networking among many nodes of the system, which in music's case include musical works, publishers and record companies, performing ensembles and venues, criticism, and scholarship. The networks that result are both the outcome of and the conditions of possibility for music's entry, around the end of the eighteenth century, into the aggregate of modern (meaning decentralized, self-regulating, and self-perpetuating) social systems. By drawing attention to processes of self-organization and self-generation shared by living and nonliving systems, Luhmann's work creates a vantage point from which human artifacts and cultural trends can be seen to exhibit formal tendencies not unlike those in other complex biological and physical settings. Exploring such similarities broadens the scope of the terms *system* and *network* beyond their customary technological domain. Indeed, tantalizing new research on forest ecology challenges the metaphorical supremacy by which the natural world is viewed through the lens of technology and expected to conform to the limitations of human understanding. Pollan describes how trees exchange chemical signals and life-giving resources by way of an underground network of mycorrhizal fungi—a kind of "wood-wide web."[11] Reversing commonplace assumptions about which creatures are the most highly evolved, biologist Stefano Mancuso states that plants, not humans, are "the great symbol of modernity." Plants, in short, help us imagine what Pollan calls a "future that will be organized around systems and technologies that are networked, decentralized, modular, reiterated, redundant—and green."[12]

Such views resonate with my own forward-looking but conservation-minded approach. That is, I wish to conserve affinities between music and the vegetal kingdom intuited by nineteenth-century listeners while transposing organicism into a register more in keeping with contemporary scientific and philosophical thought. To say this is not to claim that current perspectives on the natural world enjoy the status of incontrovertible truth; I do, however, mean to imply that knowledge can improve despite the epistemological limitations of human consciousness.[13] My goal in this chapter is not to provide a definitive explanation of what constitutes either an organism or organicism in music but to offer a fresh account of the musical features and philosophical outlooks that contributed to the critical turn toward organic metaphors in the years around 1800. This account is necessarily bound up with stylistic features of European instrumental music in the late eighteenth and early nineteenth centuries; however, I believe that music's ability to create impressions of more-than-human vitalities in the minds and bodies of its listeners is not the privilege of any one style or tradition. I also believe that, in an era when the threats to nonhuman life on this planet are too numerous to count, such impressions should be a source of wonder rather than cause for suspicion. At the very least, the legacy of organicism challenges us to think anew about what our bodies, our sociality, and our creativity share with other living entities and the ecologies in which they are enmeshed.

The Paradox of Part and Whole

What does it take to hear music, with Adorno, as if it had not been "manufactured," as if it were "growing towards its final purpose without the intervention of the subject"? The philosopher did not offer much explanation for this blurring of *physis* and techne apart from the sense of forward motion generated by chromaticism. In the twentieth century, Adorno thought, musicians had lost the ability—and, more important, the desire—to conceal their inherited materials behind a facade of inevitability. Those materials were now too stereotyped, too reified to sustain the illusion of naturalistic growth and development. Although *Tristan*'s idiom of plant-like proliferation was a thing of the past, composers could nonetheless strive to mimic organic modes of organization. "Art as an organized object," Adorno explained, "quite literally resembles the organism in the relationship which obtains between the parts and the whole. But with the growing similarity to the living organism, it gradually distances itself from the artifact which, after all, it must remain. The virtually total organization, in which every feature serves the whole and the whole on its side is constituted as the sum of the parts, points to an ideal which cannot be that of a work of art—that is to say, the ideal of a self-contained thing in itself."[14]

If artworks cannot achieve this ideal, it is not, as we shall see, because such pristine self-containment characterizes organisms rather than human artifacts. Adorno's conviction that organisms are distinguished by a special kind of relationship between parts and whole nevertheless had an illustrious history. In the *Critique of Judgment*, Kant defined the organism as a "natural purpose," by which he meant a self-maintaining entity whose existence cannot be traced to some external intention or end.[15] He explicated the point with reference to trees, which propagate via reproduction and grow by dint of an internally regulated process that converts nutrients into bodily matter. A mechanical watch, by contrast, does not generate the materials out of which it is constructed, and it is created to serve a purpose devised by an external agent (its designer). In a tree, parts and whole are interconnected such that "the maintenance of any one part depends reciprocally on the maintenance of the rest."[16] This idea gives rise to the principle adopted by so many later commentators—namely, that an organism's (or artwork's) "parts should so combine in the unity of a whole that they are reciprocally cause and effect of each other's form." "Every part," Kant continues, "not only exists *by means of* the other parts, but is thought as existing *for the sake of* the others and the whole." Such a being is both "*organized* and *self-organizing*," a formula that succinctly captures the complementary homeostatic and processual dimensions of organisms.[17]

Something is amiss in Kant's discussion, however, as anyone who has done some pruning around the yard or raked up leaves in the fall might suspect. Trees and other plants clearly lose parts without any threat to the whole, a fact Kant neglects to mention. The philosopher's elucidation of organic wholeness is doubly strange in that it refers to the practice of grafting, which combines separate organisms in a manner that confounds any easy conceptualization of the relationship between part and whole. "A bud of one tree engrafted on the twig of another," Kant writes, "produces in the alien stock a plant of its own kind, and so also a scion engrafted on a foreign stem. Hence we may regard each twig or leaf of the same tree as merely engrafted or inoculated into it, and so as an independent tree attached to another and parasitically nourished by it."[18] If this is so, then the "unity of the whole" becomes distinctly plural. Even Goethe, in his capacity as a botanist, could not remain satisfied with the idea of wholeness. In an early remark on morphology (circa 1795), he claimed, "The most perfect organism appears before us as a unified whole, discrete from all other beings."[19] A decade of botanical studies was enough to overturn his earlier judgment: "No living thing is unitary in nature; every such thing is a plurality. Even the organism which appears to us as individual exists as a collection of independent living entities."[20] Indeed, today it would be quite shortsighted to draw the boundaries of the human in a way that excluded the colonies of bacteria living symbiotically inside and on it, or to isolate plants from the fungi that help roots absorb nutrients. As Donna Haraway colorfully puts it, "Organisms are ecosystems of genomes, consortia, communities, partly digested dinners, mortal boundary formations."[21]

Strangely enough, the mutability of plants aroused a certain suspicion regarding their organic credentials. Hegel's *Philosophy of Nature* (1830), for example, initially followed Kant by defining the organism as a "totality of articulated members, so that each member is reciprocally end and means, maintains itself through the other members and in opposition to them." But Hegel held that this relational "process" results in a "simple, immediate *feeling of self*" unavailable to plants.[22] Overturning Kant's selection of the tree as the supreme embodiment of part-whole synthesis, Hegel reserved the status of organism for animals. Plants, in his view, did not display enough differentiation among parts to achieve the animal's "higher" totality. Recapitulating the argument of Goethe's *The Metamorphosis of Plants* (1790), Hegel wrote that in plants, "the *difference* of the *organic parts* is only a superficial *metamor-*

phosis and one part can easily assume the function of the other."[23] Whereas for Kant, the parts of plants were so independent as to be nearly separate entities, for Hegel they were not independent enough. Moreover, plants were too entangled with the elements of light, water, and soil to develop a dialectical sense of self. Hegel concluded that "the plant is drawn towards the outer world but without truly preserving itself in connection with what is other."[24] He thus disqualified plants from serving as a model for organic form based on autonomy and part-whole integration.

These philosophical disputes help explain why what music critics took to be organic depended on what kind of organisms they looked to for inspiration and how those organisms were understood at the time. Accordingly, aesthetic organicism in the early nineteenth century was less a coherent philosophy than a set of loosely related images, most of them botanical.[25] Organicism emerged as critics appropriated new scientific language that sought to redress the failure of mechanistic philosophy to explain the purposeful organization and intentional activity of organisms.[26] While plants offered an attractive paradigm for continuous growth, blossoming, and fruition, animals seemed better candidates for part-whole integration. These two aspects of organisms form an uneasy pair, in that one emphasizes transformation while the other is largely concerned with organization (namely, the functional distribution and operation of discrete organs within single organisms). Adorno tried to have it both ways when he recommended that composers pursue an "organic ideal" grounded in the "concrete process of a growing unity of parts and whole," as if a musical work were a hybrid of plant and animal or, perhaps, an embryo.[27]

Not surprisingly, early attempts to define music's organic qualities had trouble reconciling the competing imperatives of persistent growth and overall unity. In his famous review of Beethoven's Fifth Symphony (1810), Hoffmann contended, "Just as our aesthetic judges have often complained of a complete lack of real unity and inner coherence in Shakespeare, when only a deeper look shows the splendid tree, buds and leaves, blossom and fruit as springing from the same seed, so only a very deep penetration of the inner structure of Beethoven's music can reveal the master's high level of reflection, which is inseparable from true genius and nourished by continuing study of the art."[28] Hoffmann's conceptual struggle here is palpable, as he refers to the tree's singular origin—something no longer present to perception—in order to bind its various phases of growth into a unity. The image does not translate very well to the musical sphere. Hoffmann tried to force the Fifth Symphony's superabundance of musical ideas under the rubric of unity by proposing an essence common to them all—namely, their shared origin in a vaguely conceived "seed." Yet he admitted that this seed could not be simply equated with the symphony's opening motive, and his special pleading on behalf of the symphony's organic integration fell considerably short of the mark.[29]

Eduard Hanslick's treatise *On the Musically Beautiful* (1854) picked up on Hoffmann's image, claiming that a musical composition "develops itself in organically distinct gradations, like sumptuous blossoming from a bud.... This bud is the principal theme.... Everything in the piece is a consequence and effect of the theme, conditioned and shaped by it, controlled and fulfilled by it."[30] Evidently, Hanslick was trying to capture the sense of rightness or necessity he experienced in good musical development—the sense that each passage of music grows naturally out of something that came before. But rather than explaining how impressions of continuity can be sustained in the face of a diversity of musical materials, Hanslick evaded the problem by granting exaggerated powers of control to a single theme. Even a critic as committed to the ideal of wholeness as Adolf Bernhard Marx admitted that it was impossible to demonstrate the contribution of every element of a musical work to the whole, because "this absolute necessity, this significance that would pervade every individual detail, does not exist in any artwork."[31]

Yet does such necessity really characterize the lives, or even simply the growth, of organisms? Michael Marder's philosophy of "plant-thinking" renounces the deterministic, end-oriented conception of plant development evident in Hanslick's comparison of a musical theme to a flowering bud. Responding to Hegel's unease over the plant's "endless growth outwards," Marder notes that botanical "bad infinity" challenges the presumptions of traditional philosophy by thwarting completion and closure.[32] The sheer proliferation of plants, their production of many more seeds than can ever take root, means that the orderly sequence of germination, budding, blossoming, and bearing fruit, so often invoked to make sense of human endeavors, is realized in only a fraction of cases. Furthermore, even though plants are often treated as exemplars of a purely internal process of growth, Marder highlights that growth's "hetero-temporality"—namely, its dependence on external factors such as weather conditions or, in the case of cultivated plants, human manipulation through the use of fertilizers, chemical ripening agents, and other such tactics.[33] Marder's analysis reminds us that the autonomy of organisms must be understood in relation to their environments—a point to which I will return.[34]

Part of the challenge facing any would-be organicist discourse is that concepts such as totality, unity, and wholeness are much easier to conceive as static achievements than as ongoing processes. The familiarity of this point does not invalidate the need to reiterate it. The organization of living beings, and accordingly their wholeness, is not like that of a well-organized desktop or piece of machinery. Rather, the wholeness of organisms is continually in the process of being produced. In a recent study of the emergence of mind, Terrence Deacon praises Kant for recognizing that intrinsic finality (orientation to an end) and the capacity for self-formation are essential features of organisms.[35] But Deacon argues that the conventional image of organisms as wholes composed of parts is too simplistic. This image, he maintains, applies more readily to machines than it does to organisms.[36] That is, machines have distinct parts that are engineered separately and then assembled, whereas the "parts" of organisms, which develop through the multiplication and differentiation of cellular material, are not self-contained, independent units (despite Kant's intimations to the contrary). Deacon suggests instead that organisms are made up of processes that together serve the purpose of self-maintenance, a conception that better suits such phenomena as a tree's shedding of leaves in autumn.[37] Organisms, from this perspective, are "(w)holes," because the purpose to which they are oriented—the maintenance of life—is not something achieved once and for all, nor is it literally present in their physical substrate. This purpose is an end, but it is an end that forestalls *the* end. It is therefore misleading to refer, as Adorno did, to the final purpose of growth. Similarly, the purposive character of a musical work is expressed in its sounding, not in its ending. Works of music, no less than organisms, demand to be understood as exhibiting what philosopher Evan Thompson calls the "dynamic co-emergence" of parts and whole.[38]

One critic who seemed to appreciate this demand was Christian Friedrich Michaelis. In his 1806 meditation on the nature of music, Michaelis asserted that the "raw stuff" of tones and melodies could not simply be placed next to one another to make music. The musical art resides in form, which involves the "demarcation and combination" (*Begränzung und Vereinigung*) of raw materials. In Michaelis's words, "*Mechanical* composition still does not yield an art form. For that *organization* is necessary; that is, the tones must enter into functional, reciprocal relations with each other, must exactly suit and agree with one another. Gradation, accentuation, the division of time, rhythm, and proportion in the combination and progression of intervals lend the tones organic form."[39] Michaelis's emphasis on organization and reciprocity shows him to be a careful reader of Kant. But Goethe's botanical studies also seem to be lurking in his description of music as "change, variation, origination, growth, diminishing, fading away."[40] Michaelis realized that temporality—namely, the organic temporality of "Wachsen und Werden" (growth and becoming), in Helga de la Motte-Haber's words—had to figure into accounts of the reciprocal organization of musical elements.[41] What is more, he understood that a reductive approach consisting in the decomposition of a piece into its building blocks could never account for music as a perceptual experience. The listener's imagination, Michaelis proposed, shapes music's "organic constituents" into a whole.[42] To illustrate this point, think of the recognition of a melody, a feat that has given phenomenologists food for thought for over a century.[43] After the first few notes, one begins to hear a continuous shape, and the moment when that shape "clicks" reflects back on and draws together the notes already heard.[44] Another moment passes, and the composition of the melodic present—a present that includes generous helpings of past and future—has already changed. While a notated melody may appear to be a finished product, in performance that melody is more holistic—more "(w)hole"—than whole.

Michaelis's discussion points to the phenomenon of emergence in music, whereby, as Albert S. Bregman states, "properties of musical sound . . . emerge from perceptual integration of acoustic components over time and across the spectrum."[45] The fundamental involvement of the perceiver in musical emergence, of what Michaelis called the listener's "imagination and inner receptivity," suggests that his composer-centered concept of organization needs to be expanded to include self-organization.[46] Broadly speaking, self-organization is a dynamic process of pattern formation that occurs in both living and nonliving systems.[47] Rather than describing the intentional activities of a discrete self, self-organization refers to the emergence of global forms out of local interactions among elements, similar to the way melody as a perceptual gestalt emerges out of local interactions among tones. Phenomena such as the V-shaped formations of flying geese, the repeating hexagons of a honeycomb, and the collective signaling of fireflies all result from processes that self-organize. Patterns in organisms, such the stripes on a zebra or the regular number of petals on a species of flower, are increasingly being understood as the result of self-organizing stages of growth and development rather than as the mere execution of a genetic program.

The sheer abundance of musical patterns—many of which are based on the recurrence of self-similar gestalts, a phenomenon often witnessed in the natural world—points to complex and largely unexplored processes of self-organization in musical creativity, whose parameters include not just the intentions of composers and performers but also the constraints of particular tonal and modal systems, the periodic rhythms of organic processes and bodily motions, the material construction of instruments, and the deep history of human physical and mental aptitudes.[48] The history of music, which includes developments in modal, metrical, tonal, formal, and rhythmic organization, might profitably be rewritten from the standpoint of self-organization in cultural domains. This is not to say that cultural production assumes forms that can be reduced to those found in nonhuman settings. Rather, it is to explore how form-creating tendencies straddle inorganic, organic, and human domains and to ponder how aspects of the former are redeployed in the emergent settings of the latter.[49] While the challenges to such an approach are many, Niklas Luhmann has suggested that one path forward lies in viewing human cognitive development—for instance, the way a child learns language—as arising from self-organization among coupled systems.[50] Eugene Narmour makes a similar argument with respect to how listeners become accustomed to a musical style.[51] Style is a "replication of patterning," in Leonard Meyer's words, and patterning informs music at multiple levels and in diverse dimensions: those of pulse, meter, and scale; of intervallic collections, harmonic progressions, and formal plans; of rhythmic and melodic motives, themes, and conventions.[52]

Strictly speaking, a performance of notated music does not qualify as self-organizing since the score typically provides a top-down blueprint (if not a fully determinate one) for music's patterned sounds.[53] Yet from an acculturated listener's perspective, music performed or played back in real time might justifiably be described as a dynamic process of pattern formation whose continuously shifting gestalts are not simply caused by an external agent or agents—say, the performer, score, or composer—but emerge in a fashion peculiar to the musical art (a situation that may partly account for Adorno's observation that music can appear to unfold "without the intervention of the subject"). Nils L. Wallin develops a concept of musical works as "evolving structures" that are "self-organizing from a state with a low degree of structure" (say, at the beginning of a piece) to one with a "high degree of structure," an outcome traceable to the "transitional, non-equilibrium character of musical processes."[54] Moreover, while a performance requires energy and materials that are not contained within the score, the resulting music occupies an emergent perceptual arena whose energies are not the same as those of the performers.[55] "Emergent phenomena," writes anthropologist Eduardo Kohn, "are nested. They enjoy a level of detachment from the lower order processes out of which they arise. And yet their existence is dependent on lower-order conditions."[56] The "level of detachment" to which Kohn refers offers an alternative means of conceptualizing musical autonomy as indicating neither music's nonrepresentational character nor its independence from society—that is, as neither the "material" nor the "social" autonomy Mark Evan Bonds identifies in music-aesthetic discourse—but the distinct stratum of reality that houses music's energetic processes.[57]

The patterning of musical sounds gives rise not simply to abstract forms but to sonic analogues of gesture, movement, and affect—analogues general enough to evoke animate (and inanimate) behaviors and actions that stretch beyond the boundaries of the human, however those might be defined. Music's ability to convey liveliness owes much to the way its "organic constituents" (in Michaelis's words) constitute emergent auditory phenomena. Organicism as a compositional practice might be understood as an attempt to amplify music's emergent character such that gestalts like motives and themes exceed their local contexts and become generative of form on larger scales.[58] In other words, organicism names a kind of formal organization that aspires to the condition of self-organization—to a self-determining process of pattern formation that ranges from the smallest motives to the shape of entire movements. To create a "semblance of the organic," music's self-organization must not appear to be like that of a nonsentient thing, such as a whirlpool or a hurricane. On the contrary, organic music must seem to engage in spontaneous and adaptive (even intentional) behaviors, despite the fact that, in notated music at least, those behaviors have been largely planned out in advance. "Spontaneity amid the involuntary," Adorno wrote, "is the vital element of art."[59] Wallin also marvels at music's combination of "periodicity and aleatorics," proposing that the "dynamic dichotomy between correlation and asymmetry is similar to that which characterizes living, organismal systems."[60] Organic music imitates not the look or sounds of nature but the mode of being of Kant's organism as an "*organized* and *self-organizing being*" that is "both *cause and effect of itself.*"[61] The locus of imitation shifts, as Bonds phrases it, from the products to the process of (self-)creation.[62] Scott Burnham similarly maintains that Beethoven's music in particular came to "embody the form of artistic mimesis newly privileged in the late eighteenth century: the imitation of *natura naturans*, or the process of nature, supersedes that of *natura naturata*, the product of nature."[63]

Recent discussions of organic life have redefined the organizational features described by Kant as *autopoiesis*, a term introduced in the 1970s by cognitive biologists Humberto Maturana and Francisco Varela. In accordance with this lineage, Thompson (whose discussion of organisms I referred to earlier) adheres to a strictly biological notion of autopoiesis modeled on the cell as a "self-producing bounded molecular system."[64] However, he notes that autopoiesis is a special type of the autonomous organization that characterizes many other dynamic systems. Thompson defines a *system* as "a collection of related entities or processes that stands out from a background as a single whole, as some observer sees and conceptualizes things."[65] In performance, music consists of "related entities or processes" on multiple levels (rhythmic, melodic, harmonic, motivic, and so on) that stand out in an analogous way from a larger sensuous background. What Hanslick referred to as the "reciprocal correspondence between melody, rhythm, and harmony" in Beethoven's *Prometheus* overture is strongly redolent of Thompson's conception of the "constituent processes" of a system, processes that "recursively depend on each other for their generation and their realization as a network."[66] Any such system displays organizational closure, in that it is characterized by a "self-referential (circular and recursive) network of relations that defines the system as a unity."[67] At the same time, an autonomous system is necessarily coupled to an environment, from which it derives the energy that allows it to maintain a far-from-equilibrium state—a state whose name, in the case of organisms, is life. If one considers being-in-performance the musical state equivalent to the aliveness of an organism, then one must distinguish musical works from organisms because the former can live multiple lives. Musical works are not ideal structures but dynamic processes *in potentia*, processes capable of multiple realizations of their systemic (w)holeness.[68]

This is not to say that all instances of music can or should be considered autonomous systems, especially since some contemporary discussions of music define it as almost any sound attended to and appreciated for its acoustic qualities. As Thompson's formulation makes clear, the question of whether an organized phenomenon constitutes a system cannot be answered without also specifying an observer—or, in music's case, an acculturated listener—who apprehends it that way. This in turn suggests the need for a broader historical perspective on music at the time when organic imagery first began to circulate among critics. The following section turns to Niklas Luhmann's social systems theory in hopes of clarifying how musical works came to be understood as dynamically self-generating in a manner akin to living things.

Music Observing Itself

As his advocates frequently lament, Luhmann has received a less-than-enthusiastic reception among English speakers due to not only the abstractness and sheer volume of his writings but also what is taken to be his antihumanist outlook.[69] His theory of social systems offers a provocative alternative to standard notions of organic wholeness, though one rather different in spirit from the more biologically oriented writings of Thompson and Deacon. Also inspired by the work of Maturana and Varela, Luhmann replaced the principle of part-whole integration with multiple system-environment relations.[70] For example, although the practice of dissection promoted a view of organic bodies as assemblages of separable parts, each so-called part of the body is a meeting place for interconnected but functionally independent systems—circulatory, nervous, immune, lymphatic, and so on. These systems are operationally closed in the sense that, for example, the pulmonary system cannot assume the function of the circulatory system, yet they are also coupled so that, in this case, the lungs can supply the blood with oxygen. Human existence, argues Luhmann, arises from couplings between three types of systems: bodily, psychic, and social. He writes, "A human being may appear to himself or to an observer as a unity, but he is not a system. And it is even less possible to form a system out of a collection of human beings. Such assumptions overlook the fact that the human being cannot even observe what occurs within him as physical, chemical, and living processes."[71] The systems that converge in human beings are not so much external to one another, as assemblage theory might have it, as highly constrained in the kinds of interactions of which they are capable.[72] To further complicate matters, Luhmann did not consider systems to be objective realities; rather, they must always be defined from the standpoint of particular observers, who are themselves embedded in (and expressions of) multiple systems.

One of Luhmann's signal innovations was to argue that social systems engage in the self-generating and self-regulating behaviors normally associated with living organisms. To indicate these properties, he borrowed the term *autopoiesis* from Maturana and Varela.[73] Biological autopoiesis (as in, say, a single cell) requires self-sustaining chemical reactions, the self-production and maintenance of physical components, and the self-generation of a membrane protecting the interior of the organism from its environment.[74] In the case of social systems, the boundaries between system and environment are not material but operational in nature. Luhmann posited that modern societies differentiate themselves into autonomous systems (legal, economic, political, and so on) that can perturb or couple with one another but can never be fully coordinated or controlled by individuals, not even by presidents or chairmen of the Federal Reserve. Instead, modern social systems are self-maintaining, a trait they share with biological and psychic systems. Luhmann's theory therefore does not support visions of a social order whose parts are fully consistent with or transparent to one another. It does, however, support the analysis of "micro-social" relations and their incremental impact on the reproduction of social systems, which takes place through recursive operations confined to each system's distinct sphere.[75] In this context, recursion can be understood as an iterative but not strictly algorithmic process that produces novelty by referring back to and altering something previously manifested. For example, speech acts can only take place by referring to and building on prior speech acts. By ensuring that changes of state are dependent on prior states, recursion drives both organic and nonorganic autopoiesis. Jürgen Habermas concludes that Luhmann's theory is "metabiological" in that it represents "a thinking that starts from the 'for itself' of organic life and goes behind it—[to] the cybernetically described, basic phenomenon of the self-maintenance of self-relating systems in the face of hyper-complex environments."[76]

Music, like the other arts according to Luhmann, is a subsystem of the social system of communication. While the existence of music depends on the couplings it establishes with other systems—not least, the musicking minds and bodies of human beings—its systemic character resides in the musical sounds exchanged among composers, performers, and listeners.[77] As Luhmann remarks, "artistic communication could never come about without society, without consciousness, without life or material. But in order to determine how the autopoiesis of art is possible, one must observe the art system and treat everything else as environment."[78]

Luhmann viewed modernity as the displacement of hierarchical social arrangements, or "stratification," by a congeries of functionally independent social systems, a transformation that began in the Renaissance and culminated in the late eighteenth century. Glossing Luhmann's conception of art history, Harro Müller writes, "In modernity, works of art are no longer regulated by rhetoric, rule-bound poetics, or various conceptions of mimesis . . . they are system/environment units within the system of art oriented toward innovation and yet also always involved in copying"—copying, that is, from prior works in a way that supports the continued existence of the system.[79] Artistic evolution, from this perspective, consists of morphological changes brought about by the selection and variation of constructive elements deployed in artworks.[80] As a system differentiated from religion or politics, art becomes, in Müller's words, "autonomous" and "independent" in that it proceeds according to its own code (beautiful/ugly) and its own medium-specific operations.[81] Autonomy in this context measures a system's self-referentiality, not its independence from society. "The concept of a self-referentially closed system," Luhmann explains, "does not contradict the system's *openness to the environment*. Instead, in the self-referential mode of operation, closure is a form of broadening possible environmental contacts; closure increases, by constituting elements more capable of being determined, the complexity of the environment that is possible for the system."[82]

Luhmann's theory sheds light on the rising fortunes of instrumental music in the late eighteenth century, an idiom recognized at the time as the paragon of modern music.[83] What was new about the music of Haydn, Mozart, and Beethoven was not its transcendence of the social but the way it undermined the constraints of "stratified" artistic production—namely, the tethering of particular styles and genres to specific venues, such as the church or chamber—by threading together materials drawn from multiple generic and stylistic predecessors.[84] The stylistic diversity and semiotic density that resulted was supported by a mode of organization in which self-referential relationships within individual works multiplied, as if to compensate for the loss of the prosodic directives of a text or the dictates of functional use. These recursive relationships range from small-scale motivic recurrences to large-scale formal returns to playful manipulation of conventions. In such music, the connectivity (and effectivity) of musical operations becomes the explicit object of aesthetic interest and pleasure, as listeners learn to recognize and appreciate what James Webster refers to as the "novelty and originality" of modern musical procedures.[85]

The first movement of Haydn's "Joke" quartet (op. 33, no. 2) is a familiar example of music that features a high degree of recursion in its melodic and rhythmic discourse—music in which, as Luhmann might have put it, "self-reference" edges out "hetero-reference."[86] Unlike, say, the strict recursion of a canon, the recursivity of late eighteenth-century music serves as the basis for the "self-referential manipulation of form," which Daniel Chua identifies as the sine qua non of musical autonomy.[87] Chua argues that this autonomy is not just that of a "self-regulating system"; in his view, plenty of baroque music exhibits operational self-referentiality. Instead, he proposes, the autonomy of a piece such as the "Joke" quartet consists in the knowingness the music displays toward its own operations. Another way to put this in the language of systems theory is that late eighteenth-century music accedes to the level of "second-order observation" by appearing to observe its own observation (which means selection and variation) of prior artistic decisions, a shift that helps account for the self-consciousness that commentators routinely ascribe to this repertory.[88] For Chua, works of music conceived under this ironic regime "are not organic structures, but structures that try to *see themselves* as organic."[89] If this is so, it is also true that listeners were meant to hear the music in the same way—as analogous to a living thing.

The trick appears to have worked. The music that inspired the first organicist descriptions is just that music which exhibits the recursivity, self-referentiality, and second-order observation Luhmann attributed to the functionally differentiated system of art in modernity. In other words, late eighteenth-century instrumental music attests to a larger social process of functional differentiation even as it recapitulates that process in the aesthetic dimension. Individual works redoubled their conditions of possibility so that they themselves, and not just the larger communication system from which they emerged and which they helped establish, acquired system-like properties. The more a piece of music moves forward by referring back to and elaborating material already presented, remaking formal expectations along the way, the more it seems to enact in real time the "self-generating and self-reproducing" operations of a dynamic system.[90] At the same time, Thompson's remark that any system must be defined in relation to how "some observer sees and conceptualizes things" reminds us that apprehending a musical work as a system characterized by the dynamic coemergence of parts and whole is the act of an observer who, like the music she listens to, is embedded in social and aesthetic environments that shape expectations regarding the connectivity of musical operations.[91] In this way, music's "semblance of the organic" (as Adorno phrased it) depends not only on the internal recursiveness of individual works but also on relationships between those works, the musical environment(s) in which they thrive, and the listeners to whom they are addressed.

Consider once again act 2, scene 1 of *Tristan und Isolde*, whose music is punctuated not only by recurring leitmotivs but also by occasional reminders of the music to which the opera relates and against which it distinguishes itself. For example, Isolde's penultimate retort to Brangäne calls to mind the quickening tempo of a cabaletta (example 1.3a), while her final words to her maid begin with something very much like the *Ring*'s so-called "Flight" motive sounding in the orchestra (example 1.3b). Isolde's ode to the goddess of love culminates in a climax reminiscent of Sieglinde's "O hehrstes Wunder" from *Die Walküre* (example 1.3c), and the clamorous peroration that closes the scene features a trumpet line suspiciously similar to the tetralogy's "Curse" motive (example 1.3d). Traditional organicism might see such moments as a threat to the autonomy of individual works, but an organicism informed by systems theory would interpret them as traces of the autopoiesis of the music system.[92] Such moments, in short, complement a work's recursive relationships with recursiveness vis-à-vis other music. What Luhmann called the "self-programming" artwork must establish some such relationships if it is to remake the forms it inherits from tradition.[93]

Ultimately, then, Adorno's remark in "Vers une musique informelle" that modern music strives to realize an organic "ideal of a self-contained thing in itself" overlooks the fact that every organism requires an environment to survive and that the environment of an artwork includes not only other works but also observers. In *Opera and Drama*, Wagner went so far as to state that "the willing expectation, or expectant will of the hearer, is thus the first enabler of the artwork."[94] Luhmann portrays the matter somewhat differently in *Art as a Social System*: "The artwork ... only comes into being by virtue of its recursive networking with other works of art, with widely distributed verbal communications about art, with technically reproducible copies, exhibitions, museums, theaters, buildings, and so forth.... A work of art without other works is as impossible as an isolated communication without further com-

EXAMPLE 1.3. Wagner, *Tristan und Isolde*, act 2, scene 1: (*a*) mm. 338–49; (*b*) mm. 372–80; (*c*) mm. 384–92; and (*d*) 404–15.

munications."[95] Although every modern artwork is subject to such conditions, the modern composers who concerned Adorno in "Vers une musique informelle" strove for composition sui generis by attempting to create musical sense out of piece-specific contextual relationships. Adorno thought the "close contact between different musical complexes" realized through continuous transition accounted for the organic character of Schoenberg's atonal works, insofar as organicism makes an appearance in contemporary music at all.[96] However, defining organicism solely in terms of internal structural relations is insufficient, and not solely from the perspective of systems theory. If music is indeed an "expression of life," as Wagner once remarked, then its

EXAMPLE 1.3. (continued)

uncanny ability to suggest the animation of "trees, flowers, animals" (as Hoffmann imagined it) in addition to humans needs to be conceived in equally holistic terms.[97] Music's relationship to the other members of Hoffmann's list—stones and water—remains a question for another chapter.

From Organicism to Organism

Likening a piece of music to an organism is not, of course, the same as calling it a dynamic system, even though the two concepts do share certain characteristics. Essential to organicist rhetoric is its preoccupation with the vitality of music, with the seemingly self-sustaining but fragile animation that

EXAMPLE 1.3. (*continued*)

makes music resemble something living—something that, moreover, is not necessarily or even typically conceived as human. At this point, one cannot avoid running up against the irreducibly metaphorical or analogical (or even mythical, following Susanne Langer) nature of aesthetic organicism.[98] Perhaps it would be helpful in this regard to recall an epigrammatic remark of Hanslick's: "What in every other art is still description is in music already metaphor."[99]

The difficulty of ascertaining the epistemological status of linguistic descriptions is not, however, a problem unique to music criticism. Deconstructive approaches to metaphor have called into question the distinction between figurative and nonfigurative uses of language, a distinction that per-

EXAMPLE 1.3. (*continued*)

vades arguments about the objectivity or subjectivity of judgments.[100] "In the end," muses Luhmann, "everything is metaphorical," including his own notion of autopoiesis:

> In the sociological literature that refuses to integrate this concept into the theory of social systems, there exists the conception that it is a biological metaphor, similar to the metaphor of the organism that is applied in an uncontrolled manner and perhaps with conservative intentions with respect to social systems.... However, this is a discussion that will phase out in time, I believe. It is already an indication of little concern if someone states that something is a metaphor. If one goes back to Aristotle's *Poetics* and other traditional texts, one can say that all concepts are metaphorical.[101]

Luhmann's attitude may seem to justify a certain shirking of responsibility when it comes to deciding whether music "really is" organic, self-organizing, an autonomous dynamic system, or what have you. But such decisions may be too narrowly focused on the ontology of musical works, which is complicated by their semblance character. "How can making," Adorno asks in *Aesthetic Theory*, "bring into appearance what is not the result of making; how can what according to its own concept is not true nevertheless be true?"[102] For Adorno, these questions were intimately related to the impossibility of what was for him art's fundamental aspiration, as it was for Kant: to be like a thing of nature. As Adorno stated the paradox, "With human means art wants to realize the language of what is not human."[103] The dynamic (w)holeness and self-organization of music could only have indicated, for Adorno, the "second-order, modified existence" of natural principles in music.[104] Yet one wonders how Adorno's judgment might have differed had he entertained less rigid distinctions between humans and nonhumans, or between artistic creativity and natural productivity.

Circling back to the appropriateness of critical terminology, it should be self-evident that uses of metaphor and analogy generally do not seek to reveal objective truths but to make new perspectives and experiences possible through unexpected or surprising comparisons. This is doubly true in the case of music, since what it is and how it is perceived are not strictly distinguishable. In music studies, unfortunately, organic metaphors have largely ceased to be unexpected or surprising, and they have often been used in a manner that predetermines analytical outcomes or forecloses interpretive possibilities. James Webster, for example, decries the "organicist shibboleth of 'unity,'" which he believes obscures the messier kinds of coherence found in Haydn's "dynamic form[s]."[105] Lawrence Dreyfus complains that "figments of the organicist imagination," by which he means Heinrich Schenker's imagination, suppress compositional strategies more plausibly attributed to the intentions of composers. Even Luhmann's contention that in aesthetics, the organic metaphor has failed is based solely on the association between organicism and the principle of unity—a principle that, as I have shown, needs to be replaced with a dynamic concept of (w)holeness.[106]

Carl Dahlhaus's handling of the metaphor in *Foundations of Music History* is more circumspect, in that he sees it as a "treacherous" but indispensable resource for the historian concerned with stylistic change.[107] Dispensing with the notion of style as unfolding in stages akin to birth, growth, and decline leads, in Dahlhaus's view, to an inability to grasp transition. He gloomily sketches a dilemma in which the historian "accepts the dubious metaphysics of the organism model and its attendant normative implications, or he chooses to describe the styles of epochs in isolation—which is tantamount to dispensing with history altogether."[108] Dahlhaus's remarks bring to mind Hayden White's identification of organicism with the synecdochic method of writing history, where the phenomenon under study is viewed as a whole composed of subordinate (and subservient) parts.[109] This method, alleges Leo Treitler, instills events with an "aura of inevitability."[110] Once again, however, there is nothing inevitable about the life course of an organism—nothing other than death, that is. The vicissitudes of organic life are just as much a matter of accumulating contingencies as is the fate of a musical work or style.

In the field of historiography, Manuel DeLanda has been especially vocal in criticizing the organic metaphor's connotation of a "seamless totality" founded on "relations of interiority" conceived as separate from an environment. DeLanda seeks an alternative in a version of assemblage theory derived from the writings of Gilles Deleuze. Assemblage theory emphasizes relations of exteriority, such that any part of an assemblage "may be detached from it and plugged into a different assemblage in which its interactions are different."[111] DeLanda takes his cue from Deleuze's own biological examples—especially the symbiotic relationships between certain plants and insects—noting that "conceiving an organism as an assemblage implies that despite the tight integration between its component organs, the relations between them are not logically necessary but only contingently obligatory: a historical result of their close coevolution."[112] DeLanda's observation, rather than overturning the concept of organism, only highlights the inappropriateness of (human) logic as a measure of the complexities of nature. Contingency, again, is not foreign to the nature of organisms.

While the "conservative intentions" Luhmann mentions in regard to organicism surely deserve critique, it may be going too far to suggest that organic metaphors ought to be deployed in a controlled rather than "uncontrolled" manner. The notion of control runs counter to the rampant growth of organic (especially vegetal) life both admired and feared by critics. Although what Kant called "luxuriance" is only one of many states of vegetal being, organicism today might shed its historical ties with the principles of economy and integration and instead embrace proliferation, profuseness, even floridity in all its potential unseemliness.[113] John Daverio suggests that the organic metaphor is "well-worn" but "not yet worn out," while John Neubauer maintains that organicism is "too multifarious to be condemned wholesale."[114] My deployment of vegetal imagery in this book arises from a desire to view the products of human creativity through a less anthropocentric lens as well as to express solidarity with rather than separation from the larger realm of the living. Figuration is a venerable means by which human thought endeavors to carry traits and processes we see elsewhere in the world into the territory of our own self-understanding. To those who worry that metaphor dangerously obscures the differences between its terms of comparison, I would counter that a refusal to entertain similarities across species and kingdoms breeds a solipsistic human exceptionalism whose planetary effects are far more dangerous. As literary critic Murray Krieger has written, "I have yet to be persuaded that the organic metaphor is *necessarily* dangerous in being applied to literary texts by other critics who, as secular skeptics, retain an awareness of the figurative, and even fictive, character of that application."[115] The breed of organicism I have in mind does not merely take shelter in the safety of the fictive, however, but invites multispecies communities to flourish within the sphere of criticism, creating an ecology of the imagination founded upon the real-world implications of ecological diversity and intraorganism multiplicity.

For those unwilling to take up the banner of critical organicism, even in the revamped form I have outlined here, there remains the empirical matter of music's appeal to and effect on the human organism. In this era of fascination with the electronic technologies that support modern music making, it is worth reserving discursive space for the manifestly organic shivers, tears, sweat, ecstasy, and discomfort that music wrings from unruly and unpredictable bodies—without, however, denying that such effects may have technological (meaning, here, designed by humans) conditions of possibility.[116] As I will explore in later chapters, music influences states of physiological and emotional arousal, involves multisensory and not just auditory perception, activates the skeletal-motor system even in listeners who are still, and invites imaginary, and often unconscious, participation in its physical production.[117] Music's couplings with bodily systems ranging from the autonomic nervous system to the chemical pathways of emotion to conscious thought are as varied as the acts of performance, dancing, working, private listening, interpretation, exercising, or trancing that music accompanies. Friedemann Kawohl has written that the more biological conceptions of organicism that arose in the later nineteenth century explore just these kinds of "resonances" between coupled systems—music, the body, feeling, affect.[118] Music is organic to the extent that it facilitates such resonances, but it need not adhere to traditional notions of stylistic organicism to do so. By coupling with human organisms in such manifold ways, by charting paths though cognition, bodily awareness, and affective response that do not necessarily converge into a coherent subjectivity, music may even allow us, in Marder's words, to be "dis-organized: to live outside the totality of an organism: to be a plant."[119]

2

Formalism's Flower

Eduard Hanslick may not have been the first music critic to champion what is now known as musical formalism, but his *On the Musically Beautiful* of 1854 was without doubt the most daring and extended presentation of formalist principles to date. Hanslick's book, which appeared in fifteen editions over seventy-odd years, set out to demolish the commonly held view that the value of music lay in its representation of emotions. Hanslick denied that music—especially music lacking a text—was capable of the specificity implied by any such theory. No sure path, he argued, led from a given configuration of musical tones to the impression of tender sincerity or impassioned fury or whatever expressive quality listeners might attribute to it. Hanslick explained that such impressions rely on either analogies between musical motion and the dynamic quality of emotions or symbolic associations not actually present in the music.[1] Since it was impossible to ascribe any fixed emotional or conceptual content to music, Hanslick concluded that "the content of music is tonally moving forms" (*OMB*, 29). It followed that the only sensible music criticism, meaning (for Hanslick) criticism that aspired to objectivity, was that which explicated the composer's formation of musical material—the presentation and development of themes, the shifting relationships between melody, harmony, rhythm, and orchestration, and so on. While Hanslick granted that music was endowed with "spiritual substance," he believed this substance was inseparable from the actual notes and could in no way be illuminated by linguistic concepts or scientific disciplines such as acoustics or physiology.[2]

Hanslick's treatise laid the groundwork for formalist analysis, an approach that, as it is generally understood today, focuses on how pieces of music work in a purely technical sense. The past hundred and fifty years of musical scholarship, however, illustrate that formalism has undergone considerable transformation and that its fortunes have hardly been determined once and for all. Whereas analytical studies dating from the late nineteenth and early twentieth centuries (such as those by Heinrich Schenker and Ernst Kurth) actively promoted particular cultural values, analysts later in the century tended to avoid explicit discussions of value and meaning even as they continued to promote Hanslick's "purely musical" criteria of unity, organic development, and aesthetic autonomy.[3] By the mid-1980s, musicologist Joseph Kerman had enjoined his scholarly peers to move beyond the apparent constraints of formalism, on the one hand, and positivism, on the other, both of which had sidelined music's broader humanistic import.[4] Organicism, by that time little more than a synonym for unity in variety and developmental procedures, came under fire during same period due to its presumed allegiance with formalist methodologies. The "new musicology" of the 1990s found its guiding light not in Hanslick but in Theodor Adorno, whose formalist leanings were tempered by his thesis that musical form was a repository of social meaning. Musicologists such as Susan McClary and Lawrence Kramer set about exploring how music bore witness to contemporaneous notions of gender and the body, concerns which had barely figured in previous musicological studies.[5] Taking a broader view, Richard Taruskin charged that the "sterile formalism" of composition and analysis in the late twentieth century had consigned art music to cultural irrelevance and even imperiled its survival.[6] On the whole, however, musical formalists stood firm, retorting that the new spate of interpretive endeavors were at best loosely tethered to "the music itself" and at worst simply arbitrary.[7]

Although today plenty of musicologists and music theorists seek to reconcile inquiries into history and meaning with analytical pursuits, some have expressed suspicion of (if not outright disdain for) both analytical formalism and the hermeneutic wing of the new musicology.[8] Carolyn Abbate has emphasized music's ability to absorb or shed meanings depending on particular interpretive and experiential contexts, frustrating attempts to pin down its significance.[9] Abbate leverages music's semiotic recalcitrance in hopes of recuperating the ineffable, a concept much criticized by musicologists for the way it seems to protect music from ideological scrutiny.[10] Other critics have sought to shed analysis's self-appointed mantle of objectivity by investing formalist approaches with political aims. Martin Scherzinger has urged that the possibilities for scholarly engagement with music go beyond either "apolitical analytic practice" or "anti-analytic political practice."[11] Scherzinger worries that interpretations of music which reduce it to a representation of concurrent social realities leave no room for music to expose the limits of representation itself—a role that music, as a "discourse of the unsayable," has enjoyed at least since the late eighteenth century.[12] James Currie shares this predilection for asking not what slice of the real world music might reflect but "what manifests itself in the world only as a result of music."[13] Whereas Scherzinger recommends that scholars harness formalism for politically progressive purposes, Currie resists the impulse to make music "work" for any preconceived end whatsoever. Ultimately, Currie concludes that "the preservation of a certain gap between musicological work and politics would be more productive" than continuing to reiterate the historicist injunction to place musical works in context.[14] To this end, he recommends adopting music theory's "protective framing of the musical object from cultural practices"—the quintessential formalist gesture.[15]

With this branch of the discipline in healthy flower, musicology has in a sense come full circle, returning to—and worrying over—Hanslick's separation of formalist inquiry from the investigation of music's ties to specific social and historical settings. "Nowadays," Hanslick remarked in *On the Musically Beautiful*, "it takes real heroism to declare . . . that historical comprehension and aesthetic judgment are two different things" (*OMB*, 39–40). Arguments against hermeneutics and contextualism can be viewed as updated versions of Hanslick's ban on idea- and feeling-centered criticism. Their respective objections can be traced to the premise that music cannot signify anything—whether emotions, concepts, or concurrent social realities—reliably enough to support the quest for meaning. We appear to be left, then, with form (whose relation to politics is still up for debate) or, if we follow Abbate, with private aesthetic experiences in all their singularity.

It is time to clear a new path out of this impasse. One possibility, which I will explore in chapter 6, is to diversify what we understand by the phrase "musical meaning." Here I take a different tack by suggesting that we do not need to protect music from cultural practices (pace Currie) so much as we need to expose music—or, rather, the investigation of music—to natural behaviors. As I suggested in the previous chapter, music makes manifest dynamic processes whose formal properties span the domains of human culture and nonhuman nature. The axis along which most studies of music travel—with social meaning at one end and the autonomous realm of "music itself" at the other—excludes this wider set of resonances from consideration. Moreover, efforts (such as Scherzinger's) to combine politics and analysis have remained largely humanistic in spirit, with politics defined as contestations over "race, class, gender, sexuality, and so on." Until quite recently, the politics of nature has tended to show up on musicology's radar screen only when it overlaps with another member of Scherzinger's list, while the politics of human–nonhuman relations has made for a still rarer blip.[16] Yet the autonomy of music's formal processes, in the systems-theoretical sense put forward in chapter 1, does not rule out resemblances to other broadly distributed dynamic patterns, such as what Goethe (and later Langer) referred to as systole and diastole (contraction and expansion). Even Hanslick admitted that nature supplies the "law of duple rhythm: rise and fall, to and fro" (*OMB*, 69). Such resemblances indicate that music cannot be considered an exclusively cultural practice without impoverishing our grasp of the formal territory shared by human-generated artifacts and natural entities, such as the replication of patterns, periodicity, self-similarity across scale, and hierarchical nesting.[17] It would be absurd to maintain that such features are merely the result of intentional imitation of nature, especially in the case of music. Rather, the creative generation of regularities and variations spans human and nonhuman realms of self-actualization.

This insight has taken hold in fields ranging from philosophy to anthropology. Elizabeth Grosz contends that the production of variation in biological and cultural life follows "the same principles and processes." These related manifestations of vitality, she continues, share in the "dynamic movement of elaboration that coheres in temporal emergence."[18] Similarly, Eduardo Kohn, building on the work of Terrence Deacon, argues that form-creating or morphodynamic processes characterize both the human mind and the world of which it is part, invalidating strict distinctions between the two. Form, for Kohn, consists of "constraints on possibility that result in a certain pattern," a definition broad enough to encompass an almost limitless variety of phenomena.[19] His own "anthropology beyond the human" elucidates, among other things, the formal features held in common between the branching of the Amazon basin, the distribution of rubber trees in relation to water sources, and the social structures that arose as these resources were exploited by colonizers.

This chapter undertakes the more modest task of developing a mode of criticism attuned to both the formal resemblances between music and other worldly processes and the meanings that arise from that convergence. Although Hanslick was a resolute humanist who thought music owed nothing to nature besides raw materials and rhythm—as if that were not already conceding quite a lot—his formalism looks beyond the human in the way that it portrays music as beautiful in the untranslatable manner of leaves and flowers. The critical dilemmas spawned by this thesis can be traced back to Immanuel Kant's portrayal of the flower as emblematic of the division between aesthetic and teleological judgment, the subject of his third *Critique*. For all Hanslick's emphasis on music as an excrescence of the human spirit, *On the Musically Beautiful* intimates that music occupies an intermediate position between human artifacts and natural entities, a position also suggested by Hanslick's deployment of organic metaphors. Although his comparison of musical development to "sumptuous blossoming from a bud" may strike us today as quaint, such imagery nonetheless throws categories of culture and nature, subjectivity and objectivity, aesthetics and teleology into productive disarray (*OMB*, 81). My argument proceeds from this conceptual chaos toward a teleologically inflected aesthetics that applies to music and nature alike. The teleology I seek to discern in music is not that of human goals such as self-improvement or the pursuit of rational ends but something closer to what Kant identified as the organism's capacity for self-maintenance and self-organization (its status as a "natural purpose," in the philosopher's lingo). Expanding on material introduced in the previous chapter, I elaborate on music's participation in a more-than-human teleological order in which the ends are coextensive with the means. Along the way, I revisit Kant's and Hanslick's comparisons of music to arabesque, a decorative practice that merges formal abstraction with the simulation of vegetal energies. After pondering parallel instances of disciplined vitality in Robert Schumann's *Arabeske*, op. 18 for solo piano, I close with some Hanslick-inspired reflections on the transitory nature of music and organic life.

The Beautiful Deferred

Hanslick tailored the formalism of *On the Musically Beautiful* to an educated readership for whom faith in science had superseded revolutionary politics, and he did so by harnessing the tradition of philosophical aesthetics to the quest for objectivity.[20] "The striving for as objective as possible a scientific knowledge of things," he wrote, "must necessarily also have an impact upon the investigation of beauty" (*OMB*, 1). His treatise gained polemical traction from its opposition to the thriving Romanticism of European musical culture at the time, which placed a high premium on expression and the feelings of composers, performers, and listeners alike. That Hanslick sought to revive certain aspects of Kantian aesthetics would have struck many of his contemporaries as wrongheaded; among prominent aesthetic theorists, Kant appeared to have the least sympathy for music. In contrast to the prevailing programmatic outlook by which instrumental music was routinely understood to convey concrete ideas such as heroism or love, Hanslick held firmly to the Kantian view that music "expresses no conceptual content" whatsoever.[21] The music critic's task, accordingly, was to explicate not musical meaning but musical beauty. Yet this task was hardly a straightforward one, and it was certainly not made any easier by Kant's prior reflections on the art.

Although he seconded Kant by insisting that music's form is the locus of its aesthetic interest, Hanslick tipped his Hegelian hand by granting music spiritual import, whereas the *Critique of Judgment* had classed the art as a "play of sensations" with little cognitive or moral benefit.[22] More radically, Hanslick dispensed with Kant's account of the subjectivity of aesthetic judgments in favor of an exclusive concern with the musical object qua composition.[23] Kant noted that judgments of taste ascribe beauty to things, whether artworks or aspects of the natural world. Such ascriptions, the philosopher argued, are verbal shorthand for what is in fact a subjective experience: the feeling of pleasure that arises when contemplation of something brings the mental faculties into free play, that labile cognitive modality allergic to the application of concepts. Judgments of taste, Kant maintained, are therefore not logical. One cannot list the conditions a thing must fulfill to be beautiful, because beauty is not a concept which covers a range of instances in the way that, say, the concept "cat" applies to this tabby, that Siamese, and so on (*CJ*, 25–29, 46). For Kant, beauty is not so much in the eye of the beholder as in the peculiarly pleasurable mental stimulation brought about by exemplary artworks or products of nature.

Hanslick seemed to have trouble accepting Kant's point. "From the contemplation of beauty there may arise pleasant feelings in the contemplator," he admits, "but these have nothing to do with beauty as such" (*OMB*, 3). Instead, musical beauty has to be grounded in a fashion that renders emotional response moot. To do so, the would-be philosopher of music must adopt "the method of the natural sciences, at least to the point of attempting to get alongside the thing itself and seeking whatever among our thousandfold flickering impressions and feelings may be enduring and objective" (*OMB*, 1). The aesthetician's job is to identify and analyze what remains constant in the aesthetic object—or, as Hanslick put it later in the treatise, "to transfer the beauty of music to tonal forms" (*OMB*, 30). In this respect, the epistemology of formalism rests on an ontology that grants musical works—and their beauty—an objective character theoretically resembling that of natural entities. But if such objectivity is problematic in the case of nature, since any organism's capacity for observation is constrained by its perceptual and cognitive apparatus, it is even slipperier when it comes to artifacts, whose significance is mediated by the aptitudes and expectations of particular interpretive communities. Hanslick's conception of the musical work is akin to a system that lacks an environment, or a system that fully constrains its environment (whether listeners or performers) rather than being coupled with it, such that each is a function of the other.

In any case, Hanslick's promise to illustrate musical beauty quickly runs aground on the shores of the ineffable. "It is extraordinarily difficult to describe this specifically musical, autonomous beauty," he concedes. "Since music has no prototype in nature and expresses no conceptual content, it can be talked about only in dry technical definitions or with poetical fictions" (*OMB*, 30). Precisely because musical beauty is autonomous, because it is "self-contained and in no need of content from outside itself," language cannot provide an adequate account of it. And yet, beauty calls. Struggling to say something, anything, about the ostensible topic of his book, Hanslick locates musical beauty almost exclusively in the province of melody, perhaps to distance it from the rhythmic contributions of nature. "The beauty of a self-subsistent, simple theme," he remarks, "makes itself known in aesthetical awareness with an immediacy which permits no other explanation than the inner appropriateness of the phenomenon, the harmony of its parts, without reference to any third factor [i.e., a linguistic concept or idea]. It pleases us in itself, like the arabesque, the ornamental column, or like products of natural beauty such as leaves and flowers" (*OMB*, 32). We may break a musical work down into its component themes, but beyond that we can inquire no further: the "philosophical foundation" of musical beauty simply becomes inscrutable (*OMB*, 34–35). Critics can only illuminate what Hanslick calls the "proximate causes" of music's effects by identifying the musical elements at play in a given case. One may wonder "why Mendelssohn's numerous six-five chords and narrow diatonic themes, Spohr's chromaticisms and enharmonic relations, Auber's short, bipartite rhythms, etc. produce just these specific unequivocal impressions." But, Hanslick replies, "these questions, of course, neither psychology nor physiology can answer" (*OMB*, 33).

Hanslick also says something else, though not so directly: that musical beauty can only be defined in terms of other aesthetic experiences that do not depend on the transmission of ideas. Aside from a brief reference to the kaleidoscope, Hanslick turns primarily to the natural world (leaves and flowers) and human appropriations of nature (the arabesque and column) in hopes of explaining music's appeal. This strategy suggests that music is in some sense analogous to natural forms (and to the kaleidoscope's "play of color and shape"; *OMB*, 29) rather than an imitation of them (this position is similar to Arthur Schopenhauer's, as we will see in the next chapter).

Hanslick's comparison of musical themes to leaves and flowers and concomitant rejection of conceptual content recalls a key passage in Kant's third *Critique*. In the "Analytic of the Beautiful," Kant distinguishes between what he calls free beauty and dependent beauty. "The first," he explains, "presupposes no concept of what the object ought to be; the second does presuppose such a concept and the perfection of the object in accordance therewith"

(*CJ*, 65). Kant's distinction has caused later commentators considerable grief, because it contradicts his earlier proposal that judgments of beauty must not have recourse to concepts at all—a proposal that would seem to rule a dependent mode of beauty out of consideration.[24] Kant elaborates on the purer of the two modes by way of a botanical example: "Flowers are free natural beauties. Hardly anyone but a botanist knows what sort of a thing a flower ought to be; and even he, though recognizing in the flower the reproductive organ of the plant, pays no regard to this natural purpose if he is passing judgment on the flower by taste" (*CJ*, 65). Merely saying "that is a rose" when faced with a rose is not an aesthetic judgment, because it refers to "a definite concept of the object" under examination (*CJ*, 78). Appreciating the beauty of a rose demands that the quest for knowledge be suspended, that the kind of judgment essential to scientific induction—namely, the "faculty of thinking the particular as contained under the universal" (*CJ*, 15)—be set aside. In a pure judgment of taste, moreover, no concept of perfection ("What a perfect rose!") or ulterior motive ("What a perfect rose to present to my sweetheart!") can be involved. Kant included in the category of free beauties those natural objects and species of art which "mean nothing" and "represent nothing." These items stock a cabinet of curiosities that contains flowers, colorful birds, seashells, the decorative foliage on wallpaper (that is, arabesque), and, finally, "all music without words" (*CJ*, 66).

What is music, alone among the fine arts, doing in this group? The truth is that Kant was uncertain about just how fine music really was. As an art that "merely plays with sensations," music supplies plenty of enjoyment but little cultivation, since it contains, in Kant's opinion, meager material for reflection (*CJ*, 174). Kant's indication of an affinity between music and natural objects such as flowers based on their capacity to please but not mean seems to demote music to shockingly low aesthetic status. Among those who took pains to refute the philosopher's argument were Johann Gottfried Herder, Schopenhauer, and E. T. A. Hoffmann. Hoffmann, for his part, changed the "nothing" that music means into "everything" by proposing that instrumental music transports listeners into the realm of the infinite.[25] But perhaps the third *Critique* unwittingly does music an honor by placing it in the same category as beautiful shells and flowers. After all, Kant held aesthetic interest in nature in higher regard than the love of art; the former, he felt, arose from respect for nature as an objective reality, an attitude resembling the moral commitment to truth (*CJ*, 140–45 [section 42]).[26] It was a point in music's favor that it did not seek to represent nature and thereby engage in what the philosopher considered deception. Instead, music relates to nature by analogy, not mimesis. All art may aspire to the condition of music, in Walter Pater's memorable formulation, but music aspires to the condition of flowers.

Hanslick's borrowing of Kant's linkage between music, flowers, and the arabesque suggests that he looked to the *Critique of Judgment* for support in claiming that music pleases us by virtue of its form alone. But while the beautiful thing was for Kant an object of nonknowledge, so to speak, insofar as aesthetic appreciation depends on the suspension of cognitive modes that refer particulars to universals, Hanslick's separation of musical causes and effects, of music's effable building blocks and its ineffable beauty which "pleases us in itself," splits music into incompatible experiential and discursive dimensions. As an object of criticism, music resembles an object of knowledge (the flower as it appears to the botanist) more than an object of aesthetic enjoyment (the flower as it appears to someone who finds it beautiful). This is because Hanslick equates objective description with criticism as a whole, barring subjective judgments of taste from the sphere of learned discourse. As critics, we can talk about the "proximate causes" of music's effects all we want; we can analyze harmonic progressions, melodic construction, rhythmic character, and whatever else with all the confidence and objectivity of a botanist describing a flower. But we can say nothing about the beauty of a passage other than that it is beautiful. Musical beauty thus serves as the silent enabler of an onslaught of talk, of hermeneutic and analytical projects that, no matter how enlightening, amount to verbal sorties directed at an object that is not quite the same as the aesthetic object, at least in a Kantian reading. Hanslick's treatise thrives on the analogy between music's mode of beauty and that of nature's—both are pleasing in themselves with no further explanation needed or, indeed, possible. The kind of criticism imagined in *On the Musically Beautiful*, however, proceeds according to a very different analogy, one that models musical works on natural objects and considers them amenable to empirical description. The first mode resists epistemological endeavors, while the second fairly begs for them.

Yet when it comes down to it, Hanslick's treatise displays remarkably little interest in description, perhaps because of the strict limitations he imposed on it—or perhaps because all his bluster about objectivity was ill suited to music's distinctly un-object-like nature. Indeed, Hanslick feared (with good reason) that the "dry" technical analysis of music merely "makes a skeleton out of a flourishing body" (*OMB*, 14).[27] He recognized that beauty resides in music's life, so to speak, and that life resides in the living—in "that which is actually sounding in each piece" rather than the graphic representation of the score (*OMB*, 81). Accordingly, Hanslick's brief commentary on the opening of Beethoven's *Prometheus* overture highlights the music's dynamism, its moment-to-moment creation of sensible form:

> The tones of the first bar, following a descent of a fourth, sprinkle quickly and softly upward, repeating exactly in the second. The third and fourth bars carry the same upward motion further. The drops propelled upward by the fountain come rippling down so they may in the next four bars carry out the same figure and the same configuration. So there takes shape before the mind's ear of the listener a melodic symmetry between the first and second bars, then between these and the next two, and finally between the first four bars as a single grand arch and the corresponding arch of the following four bars. (*OMB*, 12)

Although he makes one modest reference to the ear's "delight" upon discerning a "touch of equilibrium between old and new," the object of Hanslick's judgment in this passage is less musical beauty per se than the sequence of musical operations. He speaks as though the music undertakes actions for a reason; tones that mount upward, for instance, trickle down again "so they may" (or, more literally, "in order to") embark on a repetition that creates a dynamically unfolding symmetry.[28] Such events are characterized, Hanslick thought, by a necessity approaching that of natural law. "Since the composition follows formal laws of beauty," he notes elsewhere in the treatise, "it does not improvise itself in haphazard ramblings but develops itself in organically distinct gradations, like sumptuous blossoming from a bud" (*OMB*, 81). By the standards of the *Critique of Judgment*, Hanslick's formula is strange indeed, in that it subjects the phenomenon of beauty, which compels agreement but follows no rules, to "formal laws" similar to those directing the growth of plants. By contrast, Kant maintained that the mechanistic concept of natural law was insufficient to account for self-directing, purposive processes such as organic growth. The outcome is that the analytical judgment Hanslick exercises is poised uncertainly between aesthetics and teleology, between judgments of beauty and judgments of purpose—the two competing domains of Kant's third *Critique*.

Aesthetics, Meet Teleology

While aesthetic judgment is supposed to make do with impressions of "purposiveness without purpose"—with the way beautiful things gratify our cognitive faculties in the absence of any laws or concepts governing them—teleological judgment ponders the purpose of things, the way they appear to be designed for specific ends. Kant deemed organisms "natural purposes" because they are self-creating rather than the product of someone's design. An organism, as indicated in chapter 1, is "organized and self-organizing" and "both cause and effect of itself."[29] Kant's notion of self-organization introduces a mode of causality that transcends mere mechanism, a causality that appears to work, at least in part, from the top down (this is, essentially, the Aristotelian notion of final cause). The concept of a species, for example, seems to govern the way an acorn grows into another instance of its parent oak. Although the exact features of a purple coneflower or green peafowl seem utterly contingent as far as mechanical causality is concerned, such creatures clearly grow and reproduce in a manner that suggests relatively stable ends. But since Kant was unsure how this could be, he considered the teleological principle of natural purposes to be regulative rather than constitutive, meaning that it originates not in our actual understanding of nature but in our own epistemological needs. We judge natural entities as if they have been designed with specific ends in mind because that is the only way we can make sense of them.[30]

Evan Thompson argues that Kant's notion of teleology as a merely regulative idea reflects the shortcomings of biological knowledge at the time. As a corrective to Kant's view of nature, Thompson proposes that self-organization is a constitutive feature of the natural world whose potential to create novel forms supplies at least part of the more-than-mechanical causality that Kant could not (or would not) recognize as inherent in natural processes.[31] The purposiveness of self-organization thus replaces, for Thompson, Kantian purposiveness that derives from a concept—which means that teleology must somehow cease to imply that a thing's purpose stands outside it in a separate conceptual realm. Extending Thompson's argument, I would suggest that self-organization bridges aesthetics and teleology because it accommodates both formal characteristics and intrinsic purposiveness. That is, the self-organizing behavior of organisms accounts for both the formal properties Kant admired in them (recall his litany of "free beauties") and their realization of apparently predetermined purposes (such as maturation into a particular kind of creature). The patterns displayed by organisms—say, the swirls of ornate seashells, the radial distribution of flower petals around a center, the symmetrical patches of color on a bird, or the self-similarity across scale of a tree's branches—may appear to be static, but they arise from dynamic, self-organizing processes of growth and development.[32] This means that the forms apparent in nature's "*external* aspect" are not fundamentally different from their teleological organization, as Kant supposed (*CJ*, 222).

Since works of music are not life-forms but instances of human communication bearing the traces of human design, they cannot be said to possess an intrinsic (or objective) purposiveness equal to that of natural entities. But if art, in Kant's words, "must not seem to be designed" in order to succeed, and if art "must *look* like nature, although we are conscious of it as art" (*CJ*, 149), then it must appear to be self-organizing. Hanslick's discussion of the *Prometheus* overture calls attention to such familiar musical devices as repetition, expansion, hierarchical nesting, and the production of self-similar relations—devices that have many analogues in the natural world.[33] The point is not that musical and natural phenomena manifest precisely the same forms, but that form in both music and nature arises from dynamic processes of pattern formation that produce similar results.[34] "Floral and foliate growth," writes naturalist Richard Mabey, "somehow echo the dynamic processes of our imaginations," and he argues that representations of plant forms dating back to the Paleolithic era testify to the affinities between human "patterns of thought" and vegetal "rooting, sprouting, forking, branching, twining, spiraling, leafing, flowering, [and] bearing fruit."[35] Mabey's proliferating list of verbs hints at the presence of common morphodynamical principles that, in Thompson's words, "integrate the orders of matter, life, and mind."[36]

The dependence of purposeful activity on formal constraints may be one such principle. A series of random tones or beats conveys little purpose; the cyclical patterns of meter, the intervallic and scalar constraints on melodic construction, and the recurring paths of tonal harmony imbue music with direction, and with direction, purpose.[37] Steve Larson proposes the following axiom: "experienced listeners hear tonal music as purposeful action with a dynamic field of musical forces."[38] This kind of purposefulness is different from both Kantian natural purposes (namely, the concepts living things seem to realize through maturation) and the purposiveness without purpose of art, which is not a property of artworks but the concept-free intermingling of imagination and understanding they bring about. Although musicologists have generally construed teleology in terms of music's ability to set up and attain long-range goals, a sense of purpose imbues even the simplest musical pattern or briefest passage of periodic motion.[39] Not only the longer spans of melodic unfolding or harmonic progression but also the cyclical return of the downbeat or the ecstatic repetition of a groove exhibit purposefulness, understood not as the pursuit of an end but as the pursuit of ends that make new beginnings possible.[40] This quality is not so much the product of design as it is part and parcel of musical emergence, by which local elements (such as three notes) coalesce into larger entities (such as motives) that seem to take on a life of their own. Music's purposefulness in the moment obeys a teleology of immanence rather than transcendence; any goals it attains during the course of its sounding represent compromises between end orientation and self-maintenance in the present. Similarly, organisms undertake actions that serve future ends (such as putting out buds months before they will bloom) even as their daily cycles—however varied they may be—serve the purpose of immediate survival. In its creation of a self-sustaining field of sonic action, music transposes organic persistence into an emergent acoustic realm. In this respect, music's teleological dimension is indistinguishable from its aesthetic effect, a confluence that paves the way for a critical rapprochement between Kant's two species of judgment.

Life Lines

If music conveys vitality but is not biologically alive, if it is self-organizing but also coupled with the minds and bodies of composers, performers, and listeners, then how might we understand its distinctive mode of nonliving animation? There have been numerous attempts to answer this question over the years. Although best known for his organic metaphors, Heinrich Schenker hazarded that music is "*subject* [Subjekt], just as we ourselves are subject."[41] For Ernst Kurth, music was "psyche externalized."[42] Expanding on Kurth's theory of music as a dynamic process, Victor Zuckerkandl stated that music was a manifestation of "the purely dynamic, the nonphysical, the nonmeasurable" in nature.[43] Finally, Susanne Langer claimed that the "essence of all composition—tonal or atonal, vocal or instrumental, even purely percussive, if you will—is the semblance of *organic* movement."[44] Music, she argued, possesses "vital import" rather than referential meaning, an import consisting of "the pattern of sentience—the pattern of life itself, as it is felt and directly known."[45] Life's "metabolic pattern," which encompasses such rhythms as "systole, diastole," "making, unmaking," and "crescendo, diminuendo," bestows form on "feeling, life, motion and emotion."[46] For Langer, music is the human art that best captures this all-encompassing order of vitality.

Langer's elucidation of the vital rhythms of music is broadly, but not universally, applicable. Taruskin, for instance, has argued that over the course of the twentieth century, a "geometrical" aesthetics of musical performance edged out the "vitalist" practices of the nineteenth century.[47] In addition, plenty of recent music creates the semblance of mechanical or otherwise inorganic movement, while frequently also suggesting the highly regular rhythms of physical labor or other human activities. Aside from these stylistic counterexamples, a contemporary treatment of music's vitality would need to go beyond the interpretation of formal features by considering music's physiological and psychological effects on listeners. The liveliness of music, in other words, does not reside solely in musical form. Arnie Cox argues that music invites a kind of physical empathy whereby listeners imagine *making* or *being* the music, a response that often operates cross-modally (being the music, that is, can entail being some other thing that possesses characteristics of the music).[48] The often unconscious physical and imaginative mimesis of musical actions arises from intimate but obscure couplings between music and multiple bodily and psychic systems.

As we will see in chapter 5, Hanslick left little room for such considerations, suspicious as he was of any overtly physical response to music. Instead, he grappled with the enigma of music's liveliness by way of another analogy—namely, between music and arabesque, a decorative style modeled on the curvaceous growth of vegetation. Arabesque, whose roots stretch back to the ancient world, was the subject of frequent commentary in eighteenth- and nineteenth-century art criticism.[49] Kant's reference to patterns of foliage on wallpaper as a species of "free beauty," for example, indicated that arabesque pleases the eye while conveying no meaning. Yet arabesque is not free to the extent that it takes the undisciplined shapes of plant growth and subjects them to geometrical regularization, which Kant considered the by-product of instrumental—and therefore tasteless—human purposes. "Hardly anyone will say," he writes, "that a man must have taste in order that he should find more satisfaction in a circle than in a scrawled outline" (*CJ*, 78). Feelings of contentment in the face of geometrical designs arise merely from their implicit practical usefulness. But "in pleasure gardens, room decorations, [and] all kinds of tasteful furniture"—namely, in cases where the goal is to inspire free play of the viewer's mental faculties—"regularity that shows constraint is avoided as much as possible" (*CJ*, 78). Designers who aim to please must strive to imitate nature, which is "prodigal in its variety even to luxuriance" and displays "no constraint of artificial rules" (*CJ*, 80). The aesthetic challenge of arabesque, then, is to strike a balance between the determinateness of human concepts and the freedom of living nature, between the narrow satisfactions of geometry and the irrepressible profusion of the vegetal.[50] Dynamically curving lines, complex intertwining patterns, and replication of elements such as leaves at various scales all mitigate against the "stiff regularity" that repelled Kant.

Despite his inclusion of musical "fantasies" (and, ultimately, "all music without words") in the category of free beauties, Kant did not feel that music handled this challenge particularly well. "Even the song of birds," he mused, "which we can bring under no musical rule, seems to have more freedom, and therefore more for taste, than a song of a human being which is produced in accordance with all the rules of music" (*CJ*, 80). To be sure, the tonal music with which Kant was familiar did follow certain contrapuntal and harmonic rules, but those rules generally did not predetermine melodic construction or overall form. Pieces that appeared to realize preexisting templates or schemas were precisely those that contemporary critics condemned as "mechanical" (see the discussion of Michaelis in chapter 1). Like decorative arabesque, music must not deploy such features as symmetry and self-similarity across scale in too rigid a fashion; music that is pleasing (at least in a Kantian interpretation) makes room for flexibility, spontaneity, and approximation rather than exactitude. Similarly, Langer held that the "rhythmic character of organism permeates music," a character she defined not simply as periodic motion or constant pulse but as the "preparation of a new event by the ending of a previous one." The *"rhythmic continuity"* expressed in breathing, heartbeats, and the "intensified vitality" of emotional and intellectual experience exhibits not the "stiff regularity" of a ticking clock (or the cold geometry of Hanslick's kaleidoscope) but the flexible and responsive periodicity of life.[51]

In music, the patterning characteristic of arabesque (its "semblance of rhythm," in Langer's words) is translated into actual rhythm—into the dynamic temporality of patterned sound.[52] Hanslick's own comparison of music to arabesque exuberantly portrays this transformation of visible vines into audible lines:

> We follow sweeping lines, here dipping gently, there boldly soaring, approaching and separating, corresponding curves large and small, seemingly incommensurable yet always well connected together, to every part a counterpart, a collection of small details but yet a whole. Now let us think of an arabesque not dead and static, but coming into being in continuous self-formation [*in fortwährender Selbstbildung*] before our eyes. How some lines, some robust and some delicate, pursue one another! How they ascend from a small curve to great heights and then sink back again, how they expand and contract and forever astonish the eye with their ingenious alternation of tension and repose! . . . Finally, let us think of this lively arabesque as the dynamic emanation of an artistic spirit who unceasingly pours the whole abundance of his inventiveness into the arteries of this dynamism. Does this mental impression not come close to that of music? (*OMB*, 29)

While visual arabesque translates living shapes into an image whose stasis is counteracted by the roving eye of the viewer, musical arabesque preserves the energetic and purposeful character of organic growth but relinquishes direct mimesis of the botanical. The "vegetal existence" to which arabesque refers is, in Michael Marder's formulation, "a metaphor for vivacity itself."[53] Melodic expansion and contraction, harmonic tension and repose oscillate in the flux of "self-formation," Hanslick's term for what I have been calling *self-organization*. According to Hanslick, music's capacity for self-organization is almost metabolic in nature; his reference to the "arteries of [music's] dynamism," notes Fred Everett Maus, invests the "flourishing body" of music with a circulatory system.[54] In the organic realm, metabolism is a cyclical, self-maintaining, energy-expending and energy-creating chemical process that sustains life—that in some sense *is* life. "Metabolism," writes Thompson, "is the constant regeneration of an island of form amidst a sea of matter and energy."[55] Rhythm, as the "animating principle" or "artery that carries life" to music, in Hanslick's words, serves as the primary metabolic pathway sustaining musical "islands of form." Rhythmic energy is distributed throughout the musical "body" by way of what Hanslick called the "reciprocal correspondence between melody, rhythm, and harmony" (*OMB*, 13). Every dimension of music, it seems, is drawn into an overriding metaphorical image of work-as-living-body, a rhetorical feat that gives the lie to Hanslick's claim that music has "no prototype in nature." The audible lines of music, like the visual lines of arabesque, transform the teleological surfeit of life into the aesthetic presentation of vitality.

A Self-Organizing Arabesque

Although Hanslick intended his reference to vegetal decoration to apply to features of instrumental music in general, I cannot help but pursue the analogy further in one of the few pieces to be explicitly named an arabesque: Robert Schumann's *Arabeske*, op. 18, of 1839.[56] Since opus 18 belongs to the group of intentionally amateur works Schumann composed during his brief sojourn in Vienna, the name has generally been taken as an indication of the piece's ornamental rather than serious status.[57] Indeed, the somewhat reserved melodies of the piece bear little resemblance to the "sweeping lines" of Hanslick's arabesque, but the visual appearance of the score—with its tightly woven texture, continuously repeating motives, proliferating self-similarities, and copious slurs—makes for a passable analogue to its decorative counterpart. The narrow tessitura and regular four-measure phrases of the melody might even recall the arabesque's disciplining of natural forms by geometricization. What the *Arabeske* lacks in melodic grandeur, however, it makes up for in rhythmic propulsion: the opening melody is awash in iambs, which, along with the constant pulse of sixteenth notes, propel the music forward with subdued but unstoppable force. The insistent quality of the rhythm, repeating melodic gestures, and relentless periodic oscillation of the harmony generate a sense of momentum that sustains the impression of musical "self-formation." This is not the momentum of inanimate bodies but of life being lived. As Andreas Weber writes in an essay on biosemiotics, "The motive that is expressed in the ecstatic aspect of the living . . . must be continued existence under the aspect of its success. That means flourishing, flowering."[58]

A microanalysis of the first few measures shows just how complex the network of "reciprocal correspondences" supporting musical flourishing

FORMALISM'S FLOWER   57

can be. The opening melodic motive, which runs from the pickup on G to the downbeat of measure 1, presents in miniature the basic energetic cycle of contraction-expansion (example 2.1). The melody circles around G, accumulating rhythmic and tonal energy via the common ornamental pairing acciaccatura-appoggiatura, before releasing that energy in a leap upward to the tonic pitch C. The figure G–(B)–A–G also constitutes a miniature out-and-back motion, one stacked on the unstable second-inversion dominant seventh chord unfolding in the accompanying voices. The melody's A is dissonant against this chord, but as the dissonance resolves downward to G, the melody converges with the accompanimental arpeggio. Harmonic tension and melodic resolution exist simultaneously; melody and harmony, principal

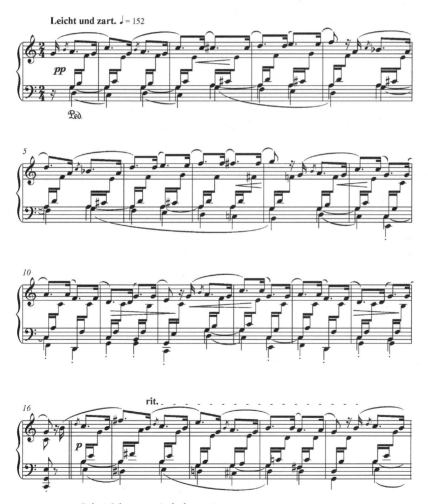

EXAMPLE 2.1. Robert Schumann, *Arabeske*, op. 18, mm. 1–20

and accompanying voices are precisely coordinated even as they exhibit different rates of change.

At measure 1, the coiled-up figure G–(B)–A–G springs upward by a fourth to C as the bass falls from D to C. These dual arrivals release energy that is metabolically channeled, so to speak, into the inner-voice sixteenths, which prepare for new events: first a repetition of the entire gesture, then a two-measure expansion (mm. 2–4). The opening measures thus establish the kind of organic regularity which for Langer consists of "the building-up of a new dynamic *Gestalt* in the very process of a former one's passing away."[59] In keeping with this dynamism, the harmonic rhythm changes on every quarter note amid the swarm of sixteenths, while the melody's relentless iambs intensify the harmonic pulsation. Multiple energetic cycles feed into one another, as the inner-voice sixteenths flow into the melody and vice versa, and melody and bass both approach C by way of rhythmically staggered whole steps (A–G, D–C). "Perhaps Schumann's genius," remark Gilles Deleuze and Félix Guattari, "is the most striking case of form being developed only for the relations of speed and slowness one materially and emotionally assigns it."[60] The basic cycle of contraction-expansion recurs on a larger scale in measures 2–4, as the melody and bass extend outward in contrary motion, like tendrils spiraling off their parent vine. The two voices traverse diminished and perfect fourths, respectively (C♯–F and G–D), infusing this venture in self-similarity—the whole passage can be heard as a wedge-like expansion of the melodic leap of a fourth across the first bar line—with a touch of asymmetry Kant might have appreciated.

Periodicities continue to proliferate, whether in the *a b a* phrase structure nested within the *Arabeske*'s main section (mm. 0–40) or in the overall rondo-like form. The pulsating main section alternates with two minor variations—the first noticeably more lyrical with its melody doubled at the octave, the second a wistful mixture of lyricism and exaggerated rhythmic articulation—both of which vary motivic material introduced in the main section. Variation I (example 2.2), for instance, starts off with the intervallically expanded version of the opening motive first heard in phrase *b* of the main section's *a b a* (example 2.1; mm. 16–17). The pace of the initial back-and-forth harmonic motion slows—doubling in the first variation, quadrupling in the second—suggesting successive relaxations of energy. The persistent "rhythmic vacillation" of the main section suggests, in its distance from vocal-style melody, not so much human expressiveness as what Marder calls the "non-conscious intentionality of vegetal life."[61] The two variations, on the other hand, evoke a more human lyricism and, in the second, even an animal impulsiveness (note the *forte* rendering of the opening flourish in mm. 144, 152, and 159; example 2.3). In this rondo, untroubled if determined botanical proliferation

EXAMPLE 2.2. Schumann, *Arabeske*, op. 18, mm. 41–48

EXAMPLE 2.3. Schumann, *Arabeske*, op. 18, mm. 144–168

gives way to duetting voices and then to restless, exploratory ambulation. The "flourishing body" of Schumann's *Arabeske* can thus be heard as a multiplicity bordering on the grotesque—that species of vegetal-animal-human design with which the arabesque often converged.[62]

Further complicating matters, two episodes in the piece fit neither the typical rondo scheme nor the variations' style of motivic work: a reflective sixteen bars wedged between the first minor variation and the return of the main section (mm. 89–103; example 2.4), and a closing section that again takes up more elongated sonorities. In these episodes, a standard process of formal unfolding (the juxtaposition of motivically related materials) gives way to a starker mode of contrast that moderates, even suspends, the forward motion of the piece. The inner-voice arpeggios slow to twice their length in the main section and become less continuous, while the downward trajectory of the melody (along with the liberal use of ritardando) effects yet another relaxation of energy. Highlighting similar episodes of self-interruption in plant growth, Marder elaborates on what he calls the "quasi-musical" rhythm of vegetal existence by way of Goethe's treatise *The Metamorphosis of Plants*, which he quotes as follows: "The organ that expanded on the stem as a leaf, assuming a variety of forms, is the same organ that now contracts in the calyx, expands again in the petal, contracts in the reproductive apparatus, only to expand finally as the fruit."[63] Marder proposes that the gaps between successive episodes of expansion in the life of plants complicate the temporal continuity and seamless development so often associated with botanical forms. From this perspective, the organic quality of Schumann's *Arabeske* is more than a matter of the motivic relationships between sections (note the fourfold appearance of the grace note–dotted eighth–sixteenth note rhythm in example 2.4). The piece's moments of contraction can also be heard as analogous to stages of a life process—to episodes when energy is conserved and concentrated before a new round of growth begins.[64]

But this plant-like expansion and contraction does not seem to be the whole story either, if the coda is any indication (mm. 209–224; example 2.5). These sixteen measures, with their slow, descending half notes and eighth-note arpeggios, clearly recall the prior episode of contraction following variation I. Yet the withholding of the main section's characteristic dotted motive until the final cadence distances this passage even more from the expressive and thematic core of the piece. The minimalistic material seems lost in reverie until it produces something altogether unexpected. In a moment of heightened rhetoric, the *Arabeske* puts forth a single flower, so to speak—one that traces its origin to a *different piece*. The little turn figure embedded in the ii6/5–vii7/V–V6/4 progression that begins in measure 214 comes straight out

EXAMPLE 2.4. Schumann, *Arabeske*, op. 18, mm. 87–105

of "Der Dichter Spricht" ("The Poet Speaks"), the final movement of *Kinderszenen*, op. 15, a piece Schumann had composed the year before (example 2.6). It is a moment in which "the freedom of the imagination is pushed almost to the grotesque," to borrow Kant's words, a moment of liberation from "every constraint of rule."[65] Perhaps, as in the grotesque, what emerges here is neither flower nor fruit but head—the head of the poet.[66]

In *Kinderszenen*, the title "Der Dichter Spricht" marks a shift in perspective that foregrounds the movement's self-reflexivity while pushing the previous movements further into the imagined past of childhood, as though their scenes have been recounted rather than reenacted. The *Arabeske*'s brief reference to "Der Dichter Spricht" similarly, if more fleetingly, represents a sudden irruption of self-consciousness into what otherwise might seem a piece lacking in aspiration (the coda's heading, *Zum Schluß* [In Conclusion], shifts op. 18's point of reference from the visual arts to literature, as Arnfried Edler notes).[67] Although this emergent self-consciousness can only be appreciated by an interpreter familiar with Schumann's other music, it does not necessarily call for hermeneutic interpretation. Instead, its significance might be

EXAMPLE 2.5. Schumann, *Arabeske*, op. 18, mm. 209–224

EXAMPLE 2.6. Schumann, "Der Dichter Spricht," from *Kinderszenen*, op. 15, mm. 1–4

construed in formalist terms as a moment when the piece seems to step out of—and redefine—its own formal constraints. That this moment is nested within the already novel coda accords with the way that the creative outcomes characteristic of emergence are not predictable from their conditions of possibility; patterns that result from advancing stages of self-organization do not obey familiar causal laws. Here one might adopt a Luhmannian perspective and suggest that the allusion to *Kinderszenen* within the *Arabeske* directs attention to the reciprocal relationship between self-organizing works and the self-maintaining (or autopoietic) system of musical production as a whole. If music written with an eye toward formal innovation observes itself observing other music and asks its listeners to do the same, as Luhmann's theory of modern artistic production stipulates, then the end of the *Arabeske* might just be an example of *third-order* observation, in which the music observes itself observing itself observing other music.

Schumann's opus 18, in sum, unexpectedly expands the scope of motivic interconnection to include other pieces, taking us far beyond what is normally considered the purview of organic relations. As an insider's moment, so to speak, the intertextual reference has to be observed in order to signify to its fullest potential, a situation that reaffirms the need to redefine the conventional formalist distinction between the "inside" and "outside" of a work in terms of system/environment relations. In addition to establishing a venue for self-referential internal relations, works must also specify a domain of interactions with the environment—a "how" along with a "what"—so they may live, and live again.[68] Schumann's *Arabeske* is addressed to a listener who can say, "Yes, I recognize that!" Music's autonomous beauty, like the beauty of leaves and flowers, may not, then, be entirely meaningless. Weber writes that autopoiesis, which for him means a mode of autonomy specific to organisms, is "the materially successful gesture of the organism's saying 'yes!' to itself." "Life as self-approval," Weber continues, "may be automatically life as beauty. In living, life celebrates itself."[69] Beauty calls, and what it says is, "I am alive!" A formalism that heeded *that* call, a formalism motivated by what Elaine Scarry refers to as the "reciprocal feeling of aliveness" inspired by beautiful things, would be far from sterile.[70] Instead, it would flourish as the afterecho of beauty, as one more affirmative stage in the "realization of the living."[71]

On Transience

But nothing lives forever. Organisms, whose purpose is to stay alive, must eventually die. While one cannot kill an artifact, one can forget it. The passage of time and inevitability of change ensure that human artifacts always shed ties with their original environments while (perhaps) forging new ones later on. At first, it seems that, for Hanslick, beauty serves as a bulwark against such change, since he claims that its enabling conditions (what he calls music's "formal laws of beauty") are permanently valid: "All musical elements have mysterious bonds and affinities among themselves, determined by natural laws. These, imperceptibly regulating rhythm, melody, and harmony, require obedience from human music, and they stamp as caprice and ugliness every noncompliant relationship. They reside, though not in a manner open to scientific investigation, instinctively in every cultivated ear, which accordingly perceives the organic, rational coherence of a group of tones, or its absurdity and unnaturalness, by mere contemplation, with no concept as its criterion or *tertium comparationis*" (*OMB*, 31). Hanslick's words portray music as a series of intractable conundrums. Musical sensibility is instinctive yet subject to cultivation; musical beauty obeys natural laws but resists scientific inquiry; music is rationally coherent but hostile to concepts. In addition, Hanslick maintained that music was exclusively a product of the human spirit, but he also held that its laws were not entirely a human invention. Instead, music sits at the intersection between organic and inorganic nature as an art form mediated by "fundamental laws of nature governing both the human organism and the external manifestations of sound" (*OMB*, 30).

Yet beauty, ever the irritant to objectivity, cannot be captured and pinned to a specimen board so easily. In a remarkably clearheaded assessment of the vagaries of aesthetic experience, Hanslick mused, "There is no art which wears out so many forms so quickly as music. Modulations, cadences, intervallic and harmonic progressions all in this manner go stale in fifty, nay, thirty years. . . . Without inaccuracy we may say, of many compositions which were outstanding in their own day, that once upon a time they were beautiful" (*OMB*, 35). Hanslick's goal of an objectivist aesthetics that would "transfer the beauty of music to tonal forms" recedes into the distance, just like the once-beautiful pieces he contemplates (*OMB*, 30).[72] Hanslick thought it was possible to identify the "proximate causes" of musical beauty using the tools of analysis, but it turns out that the relationship between musical form and musical beauty is after all not causal—not lawful at all, but contingent and subject to change.

The project of *On the Musically Beautiful* becomes even more strained as Hanslick positions himself against naïve transcendentalists who claim that great artworks persevere for all time. "We must renounce our belief," he states, "in the deathlessness of the beautiful" (*OMB*, 41). In a later edition of the treatise, published in 1891, Hanslick turned to another of his books (*Die Moderne Oper* [*Modern Opera*] of 1875) for support: "The well-known saying that the 'truly beautiful' can never lose its charm, even after a long time, is for music little more than a pretty figure of speech. (And anyway, who is to be the judge of what is 'truly beautiful'?) Music is like nature, which every autumn lets a whole world of flowers fall into decay, out of which arise new flowerings. All music is the work of humans, product of a particular individuality, time, culture, and is for this reason permeated with mortal elements of various life-expectancies" (*OMB*, 40). In a single stroke, Hanslick undercuts his own authority on matters of musical beauty—who is to be the judge, anyway? Despite attempting to cordon music off from the natural world, Hanslick admits that the beauty of this human art decomposes as readily as a flower. Beauty, then, is organic in its very capacity to fade away; it is part of the life cycle of both artifacts and natural entities. In the case of forgotten artifacts, revival and reconstruction may succeed in resuscitating their beauty, but not as a rule, and not forever. Musical beauty resembles that of leaves and flowers because it is, in its own way, a living thing, meaning that it also passes away. Music's devotees may regret this passing and blame the "evil spirit of the time," Hanslick says, but "time itself is a spirit, and it produces its own embodiment" (*OMB*, 40).[73] For this critic in philosopher's clothing, beauty is after all neither ideal nor transcendent but embodied and mortal. Faced with the wildly variable nature of music's subjective impressions, Hanslick countered with a notion of formalist objectivity that clings to something even if it is not *the* thing. Yet his treatise contains the seeds of a formalism that values music not as something accomplished but as an accomplishing, not as something embalmed by analysis but as a medium in which liveliness can be reanimated and experienced anew, at least for a time. This is a formalism captivated by music's appearance of self-organization and its acoustic manifestation of dynamic processes and patterns replete with analogues in the nonhuman natural world. This formalism, organic rather than Platonic, finds meaning in beauty for no other reason than that it is here, now. The flower will soon wilt, the music someday go stale. Enjoy it while it lasts.

3

Schopenhauer's Musical Ecology

The artistic types who populate the tales of E. T. A. Hoffmann often enjoy a degree of rapport with the vegetal world that modern readers may find exaggerated. The old composer in "Ritter Gluck," for example, describes a dream in which he hears flowers singing in an enchanted valley, while Anselmus of "The Golden Pot" is mesmerized by the whispering of elder blossoms. These quintessentially Romantic feats of sensation belie the conventional view of plants as essentially silent creatures, their only sounds being those made by wind, rain, and animals as they move through branches, leaves, and fronds. The preternatural perceptions of Hoffmann's characters, to which we will return in chapter 4, find an unexpected echo in the world of contemporary science. Recent work in plant bioacoustics has shown that plants emit sounds at low and ultrasonic frequencies, some (but not all) of which can be traced to the movement of water through their tissues. Monica Gagliano, a leading researcher in this area, has enjoined her fellow scientists to analyze these "green symphonies" for what they might reveal about plant physiology and the communicative power of vegetal vibrations.[1]

Gagliano's efforts to popularize her findings, however, have attracted criticism. In an educational video dating from 2013, she attaches electrodes to the leaves and roots of plants, feeds the signals into a custom-designed MIDI device, and encourages an audience of children and adults to listen to the "voices" of plants "singing."[2] Instead of the otherworldly timbres and random tones one might expect from such an experiment, we hear a synthesized, vaguely new age swirl of pitches drawn from the familiar Western chromatic scale. Gagliano does not fully explain what acts of transposition and autotuning make possible such a recognizably musical outcome. In a blog post commenting on the video, Michael Marder (the philosopher of the vegetal we encountered in chapter 1) charges Gagliano with "robbing plants of their own voices" in her zeal to make audible what would otherwise remain silent to human ears. As a thinker concerned with the profound differences between plants and humans as well as their points of connection, Marder complains that the kind of technological mediation performed by Gagliano does little more than impose "alien frames of reference on a given form of life."[3] He places the biologist among a long line of thinkers who have sought to translate—illegitimately, in his view—the sonic emissions of nonhuman others into concepts and categories (and musical scales) drawn from human cultural production. At the top of Marder's list of offenders sits Arthur Schopenhauer.

Had one of Schopenhauer's contemporaries asked him what he thought plants sounded like, he might have answered, a little facetiously, "tenors." As a flutist and inveterate music lover, Schopenhauer was convinced of the deep affinity between human music and the natural world. "In some sense," he muses in *The World as Will and Representation* (1819), "music must be related to the world as the depiction to the thing depicted." But this relation, he says, is "very obscure" ("sehr tief verborgen"), because music does not represent anything that exists in a nonmusical fashion, as a painting might re-create an image of the sea.[4] Music does not so much represent the world, Schopenhauer argues, as constitute an analogy of it—though attentive readers will notice that the philosopher cannot fully eliminate the vocabulary of mimesis from his discussion. Extrapolating from the varying degrees of complexity found in the voices of musical polyphony, Schopenhauer proposes that the "deepest tones of harmony" remind him of "the lowest grades of the will's objectification, inorganic nature, the mass of the planet" (*WWR*, 1:258). Continuing along this imaginary tessitura, he writes, "Between the bass and the leading voice singing the melody, I recognize the whole gradation of the Ideas in which the will objectifies itself. Those nearer to the bass are the lower of those grades, namely the still inorganic bodies manifesting themselves, however, in many ways. Those that are higher represent to me the plant and animal world." Soaring above the vegetal tenors and animalistic altos is the main melody: "In the high, singing, principal voice, leading the whole and progressing with unrestrained freedom . . . I recognize the highest grade of the will's objectification, the intellectual life and endeavor of man" (*WWR*, 1:259). Schopenhauer's analogy culminates in the startling conclusion that "we can regard the phenomenal world, or nature, and music as two different expressions of the same thing" (*WWR*, 1:262)—that thing being will, the elusive centerpiece of the philosopher's ontology.

Recent literature on Schopenhauer has largely written off his analogy between music and the natural world as either a curiosity or too narrow in its view of musical texture. Lawrence Ferrara deems the analogy no more than "myth or lore," while Lydia Goehr mentions its sources in ancient Eastern and Western cosmologies before moving on to more promising philosophical waters.[5] In terms of the history of philosophy, Schopenhauer's four-voiced image of the world can be understood as yet another variation on the theme of the Great Chain of Being, and an unusually atheistic one at that.[6] Yet cosmological interpretations of Schopenhauer's analogy, which revolve around his hypothesis that the world is comprised of entities whose harmonious relations resemble those of music, offer little in the way of reflection on what music's many voices might have in common with the ponderous matter of the earth, the tentative groping of vegetal creepers, the purposeful rooting around of animals, and the airy ruminations of humans. The multiplicity that characterizes this musical ecology, moreover, precisely mirrors that which Schopenhauer imputed to human bodies. Ultimately, Schopenhauer indicates that music affords listeners aesthetic access to organic and inorganic dimensions of existence which both inform and extend beyond the human.

This claim raises the question of what Schopenhauer's philosophy has to offer contemporary movements such as vital materialism, which, as political theorist Jane Bennett explains, seek to articulate not only the "vibrant" lives of nonhumans, but also "the 'it' inside the 'I.'"[7] *The World as Will and Representation* is remarkably catholic in its attribution of will to entities both living and nonliving, though the treatise does stop short of imputing life to the inorganic realm.[8] Could Schopenhauer, despite his reputation as a metaphysician, be a fruitful resource for critics seeking to develop less anthropocentric perspectives on human–nonhuman relations? Rüdiger Safranski gestures in this direction when he observes that, long before Darwin and Freud, Schopenhauer perpetrated the "three great affronts to human megalomania" now associated with modernity: the cosmological affront, which decentered the earth's position in the universe; the biological affront, which made man an animal like any other; and the psychological affront, which demoted the importance of conscious reflection in the totality of psychic life.[9] Humans, in Schopenhauer's estimation, may be the most highly developed expressions of will on the planet, but their physiology largely resembles that of animals and even plants. "It must not be assumed," the philosopher wrote in the second volume of his magnum opus, "that man is . . . radically different from the rest of the beings and things in nature" (*WWR*, 2:174).[10]

As this quotation suggests, Schopenhauer was unusually attuned to what humans share with other creatures, not just to what sets them apart. His theory that organic and inorganic entities alike are manifestations of will and that mineral, vegetal, and animal grades of will are present in human bodies makes his philosophy well worth revisiting at a time of burgeoning interest in more plural conceptions of human existence. Indeed, the philosopher's passion for what he called the "great multiplicity and diversity" of nature makes for a treatise whose pages are populated not only by philosophical speculations but also by crystals, alkalis, flowers, birds, beavers, marmots, bees, and spiders, among many other creatures (*WWR*, 1:153). At the same time, as the product of a "transitional thinker" with competing metaphysical and empirical commitments, Schopenhauer's philosophy—especially its perplexing rehabilitation of the Platonic Ideas—cannot simply be resuscitated wholesale. Schopenhauer's claims, on the contrary, are often frustratingly paradoxical.[11] In particular, the philosopher's notion that (some) humans can attain the status of "pure" subjects of knowing released from the imperatives of the will manifestly contradicts his eloquent demonstration of the identity of will and body. Due to its inherent monism, however, Schopenhauer's immanent rather than transcendent metaphysics harbors numerous points of contact with the many new materialisms in circulation today (*WWR*, 2:183).[12] After all, what is metaphysical for Schopenhauer—meaning will, and will alone—is that which has no prior cause, not something that exists beyond or apart from the physical.[13]

Music, Schopenhauer thought, lets us listen in on the grand pageant of will as it differentiates itself into inorganic, vegetal, animal, and human grades of being. How one responds to Schopenhauer's distinctive brand of musical naturalism depends on how one views the trope of analogy. For Marder, Schopenhauer's analogy transgresses against the specificity of worldly things in the same way that Gagliano's translation of plant sounds into equal-tempered melodies foists humanness onto vegetation. From this perspective, Schopenhauer's proposal that music and world are analogous expressions of will merely washes out the differences between cultural products and natural entities such as orchids, maples, fungi, and slime molds, not to mention rocks, rivers, meteors, and distant stars. In another meditation on a Schopenhauerian theme, Marder urges us to acknowledge the limits of empathy when it comes to vegetal creatures whose physiological variance from us entails practically inconceivable differences in embodied experience.[14]

Rather than preserving difference at all costs, however—a maneuver that, as I argue elsewhere in this book, leads into the cul-de-sac of human exceptionalism—I take Schopenhauer's musicalization of the world as a provocation, one that impels us to consider how the analogical thinking of Romanticism might be harnessed to contemporary efforts to think the human together with the nonhuman. Music appeals to (and also offends) aspects of our bodies and subjectivities not governed by conscious thought, which, in Schopenhauer's thinking, means that music acts on grades of our being that overlap with those of living and nonliving others. For all that he pined for transcendence of the body, Schopenhauer's remarks on corporeal existence dovetail with posthumanist accounts of the multiplicity of human embodiment. And, despite his focus on the metaphysical significance of the aesthetic, he drops a number of tantalizing hints about the body's intimate involvement in musical listening. In what follows, I argue that music, in an alternative formulation of Schopenhauer's philosophy, promises not so much an escape from the will as an education in the broadly ecological scope of the will's physical manifestations—with *will* understood primarily as the dynamic tendencies of matter across physical, chemical, and biological scales. By activating and engaging multiple physiological registers of listening bodies, music clears the way toward a more-than-human "ecology of self" spanning material, organic, and psychic domains.[15]

The Identity of Body and Will

As John E. Atwell has written, "It would not be an exaggeration to dub Schopenhauer *the* philosopher of the body."[16] Indeed, few philosophers rival Schopenhauer when it comes to portraying the ins and outs of embodiment, not to mention the peculiar pathos of being a living creature in a world marred by suffering and violence. Nor did his philosophical contemporaries value as highly what finely tuned observation of the world and its occupants, both organic and inorganic, had to offer epistemology. "*Empirical consciousness*," wrote this erstwhile purveyor of metaphysics, "alone is the immediate, the actually given." "Every philosophy," he continues, which "takes as its starting-point arbitrarily chosen abstract concepts such as, for example, the absolute, absolute substance, God, infinite, finite, absolute identity, being, essence, and so on, floats in the air without any support, and so can never lead to a real result" (*WWR*, 2:82–83). Schopenhauer instead intended his philosophy to begin with what he witnessed in the natural world, along with the special insight into reality he believed embodiment afforded.

While his studies of anatomy and physiology at the University of Göttingen (1810–11) stoked his fascination with the body's empirical characteristics, Schopenhauer also viewed the body as a unique source of knowledge, one that opens onto the wider, indeed universal, realm of will as the noumenal aspect or "in-itself" of all phenomena.[17] "My body and my will are one," Schopenhauer states categorically in *The World as Will and Representation* (*WWR*, 1:102). Will and body, in other words, are not two sides of a dualism but dual aspects of a monism. What I see when I look at my body, or learn about it through the sciences, or recognize as its enculturation, corresponds to its phenomenal aspect, to body-as-appearance. But I am also aware of my body from the inside, in acts of somatic introspection, so to speak. In this sense, I experience my body as will "in so far as I am conscious of it in an entirely different way comparable with no other" (*WWR*, 1:103). Humans are most aware of the "endless striving" of the will in the form of immediate physical needs, whose gratification serves a single purpose—survival (*WWR*, 1:164). No other rational principle can be found to explain why we expend so much "vital energy" in the pursuit of life, when what we get in return are fleeting pleasures along with "want, misery, trouble, pain, anxiety, and . . . boredom." Our high-minded efforts at self-realization are driven by no more (and no less) than the "same thing that makes the plant grow" (*WWR*, 2:359). The will, which orchestrates human lives just as it does those of bats, bottlebrush trees, and bacterial colonies, is the prime mover behind every physiological and psychological process, whether conscious or unconscious, voluntary or involuntary. Even the intellect, Schopenhauer contends, is merely the servant of will. He attempts to make an exception for allegedly "pure" states of knowing, such as those he imputes to aesthetic contemplation, but his reasoning on this point, as we will see, is less than convincing.

Schopenhauer calls the identity of will and body a "philosophical truth" exempt from the principle of sufficient reason, which means it has no explanation like those available in the natural sciences, logic, geometry and math, and the analysis of human motivations (*WWR*, 1:102). The truth of the identity of will and body is arrived at by intuition, as is the conviction that everything in nature is an expression of will. Intuition, in this case, proceeds by analogy: it is one's "double knowledge" of the body as object of study and site of introspection that allows one to ascribe the same "inner being" to all other creatures and natural phenomena (*WWR*, 1:104–5). Our bodies, in other words, are the key to knowledge of others: "We shall judge all objects which are not our own body," Schopenhauer states, "according to the analogy of this body" (*WWR*, 1:105). The philosopher therefore expects the reader awakened to the murmurings of his will to "recognize that same will not only in those phenomena that are quite similar to his own, in men and animals, as their innermost nature, but continued reflection will lead him to recognize the force that shoots and vegetates in the plant, indeed the force by which the crystal is formed, the force that turns the magnet to the North Pole . . . all these he will recognize as different only in the phenomenon, but the same according to their inner nature" (*WWR*, 1:109–10). The analogical thinking that proceeds from one's own somatic awareness makes possible a kind of imaginative empathy in which distinctions between humans and nonhumans recede, revealing traces of will in all things.[18] Incidentally, it is along the same pathway of intuitive recognition that listeners come to interpret musical movement as the movement of will.

The metaphysical thrust of Schopenhauer's thought, his conviction that everything shares in the same essence, is undeniable. One might wonder, indeed, what kind of purchase on reality something so general as "the will" really provides, given that it manifests itself in everything. At the same time, Schopenhauer's will-based monism is decidedly nonreductionist. That is, while all phenomena are expressions of will, the particular way in which will imbues each "grade" of material existence cannot be explained by appealing to some more primitive layer. As Barbara Hannan suggests, we can consider Schopenhauer an early proponent of emergence, or the hypothesis that "different sorts of property or causal power," such as the capacity for self-repair found in living things, "appear at different ontological levels."[19] Objecting to what he called the "false reduction of original natural forces to each other," Schopenhauer insisted that biological processes could not be reduced to chemical ones, nor chemical ones to physical forces such as gravity or magnetism (*WWR*, 1:123).[20] At the same time, every organism is made up of matter that is subject to physical forces and possesses identifiable chemical elements—hydrogen, oxygen, carbon, potassium, and so on. What this means is that all living bodies, including human bodies, are palimpsests of will, or multiplicities that routinely unsettle the seamless integration normally attributed to organisms (see chapter 1).[21] As Schopenhauer explains, "every phenomenon of the will, and even that which manifests itself in the human organism, keeps up a permanent struggle against the many chemical and physical forces that, as lower Ideas, have a prior right to that matter" (*WWR*, 1:146).

Here as elsewhere in *The World as Will and Representation*, Schopenhauer's recourse to the Platonic Ideas throws a wrench into his discussion of embodiment. In brief, the Platonic Ideas signify the gamut of forms in which the will objectifies itself. Schopenhauer describes the Ideas as "the definite species, or the original unchanging forms and properties of all natural bodies, whether organic or inorganic, as well as the universal forces that reveal themselves according to natural laws" (*WWR*, 1:169). Although this sounds like a recipe for an implausibly static universe, Schopenhauer also maintains that "each more highly organized state of matter succeeded in time a cruder state. Thus animals existed before men, fishes before land animals, plants before fishes, and the inorganic before that which is organic" (*WWR*, 1:30; this litany is probably more a holdover from the biblical story of creation than an anticipation of Darwin's theory of evolution). He seems not to have realized that if this were true, some kind of gradual change of organic forms must have been necessary; otherwise, new creatures would simply have had to pop into existence out of nothing. Not surprisingly, Schopenhauer had trouble reconciling the stable identity that the Platonic Ideas imply—as in the Idea of the human, the vole, or the baobab—with his depiction of living creatures as forums for the internal struggle between Ideas corresponding to grades of objective existence. His solution was to equate the Idea of a creature with just that part of the whole (or, following chapter 1, the "(w)hole") that is unique to it. This Idea-as-identity comes in and out of focus as it vies for supremacy with less differentiated Ideas:

> Hence the comfortable feeling of health which expresses the victory of the Idea of the organism, conscious of itself, over the physical and chemical laws which originally controlled the humors of the body. Yet this comfortable feeling is so often interrupted, and in fact is always accompanied by a greater or lesser amount of discomfort, resulting from the resistance of those forces; through such discomfort the vegetative part of our life is constantly associated with a slight pain. Thus digestion depresses all the animal functions, because it claims the whole vital force for overcoming by assimilation the chemical forces of nature. Hence also generally the burden of physical life, the necessity of sleep, and ultimately of death; for at last, favored by circumstances, those subdued forces of nature win back from the organism, wearied even by constant victory, the matter snatched from them, and attain to the unimpeded expression of their being. It can therefore be said that every organism represents the Idea of which it is the image or copy, only after deduction of that part of its force which is expended in overcoming the lower Ideas that strive with it for the matter. (*WWR*, 1:146)

Schopenhauer's attempt to skim the Idea of an organism off the larger, multiply graded plurality he so elegantly depicts seems to serve the purpose of grounding the apparent stability of species in a quasi-metaphysical typology—with the Ideas understood as an ontological way station between undifferentiated will and individuated, real-world objects—in lieu of a more materialist account (say, one rooted in genetics). The lopsided notion of identity this typology produces, however, does not mesh very well with the philosopher's more down-to-earth insights regarding the multilayered nature of human (and nonhuman) embodiment. Schopenhauer points out that not only are we made up of matter that decomposes after death (a victory for the "chemical forces of nature" described in the extracted quotation), but also the human capacity for rational thought, the defining Idea of the species, as it were, sits atop a deep sea of unconscious and involuntary organic processes. In volume 2 of *The World as Will and Representation*, he elaborates, "Let us compare our consciousness to a sheet of water of some depth. Then the distinctly conscious ideas are merely the surface; on the other hand, the mass of the water is the indistinct, the feelings, the after-sensation of perceptions and intuitions and what is experienced in general, mingled with the disposition of our own will that is the kernel of our inner nature" (*WWR*, 2:135).

If there really is a submerged kernel of human uniqueness, it is nevertheless enclosed, even overwhelmed, by the fleshy fruit of unconscious and involuntary bodily processes. In his study *On the Will in Nature*, Schopenhauer describes a handful of these processes in graphic detail: "The quickened pulse in joy or fear, blushing in the case of shame or confusion, the pallor in terror, also in concealed anger, weeping over sorrow, and erection from voluptuous thoughts, breathing difficulty and accelerated intestinal activity in the case of great anxiety, watering of the mouth at the sight of appetizing dishes, nausea from seeing disgusting objects, violently accelerated blood circulation, and even an altered quality of the bile through anger, and of the saliva through intense rage."[22] This corporeal cacophony of will comprises the accompaniment to our daily lives, if normally in a more muted register. Whatever solo is reserved for thought in its pristine rationality—the only truly native aspect of Schopenhauer's Idea of the human—is only one voice, however incessant and self-important it may be, among the resonant bodily din.

Occupying an even deeper stratum than these involuntary sensations and affects is the "blind will" to survive, the "vital energy" that keeps humans on the move just as it keeps the leaves of a houseplant directed toward the light. The more Schopenhauer contemplates the sovereignty of the instinct to survive, the more human existence comes to resemble that of our distant vegetal cousins:

> Unconsciousness is the original and natural condition of all things, and therefore is also the basis from which, in particular species of beings, consciousness appears as their highest efflorescence; and for this reason, even then unconsciousness still always predominates. . . . Plants have at most an extremely feeble analogue of consciousness, the lowest animals merely a faint gleam of it. But even after it has ascended through the whole series of animals up to man and his faculty of reason, the unconsciousness of the plant, from which it started, still always remains the foundation, and this is to be observed in the necessity for sleep as well as in all the essential and great imperfections, here described, of every intellect produced through physiological functions. And of any other intellect we have no conception. (*WWR*, 2:142)

In the context of nineteenth-century philosophy, it is hard to imagine a clearer statement of the emergent character of mind (as the "intellect produced through physiological functions") or of the persistence of the plant-like in the human.

We are now able to understand the access Schopenhauer believed we have to the "inner nature" of all things as a product of not only the metaphysical unity of the will but also the material plurality of our own embodied natures. We ourselves are hybrid creatures, or multiplicities composed of human, animal, plant, microbial, and inorganic strata, each of which exhibits varying degrees of differentiation and complexity. Although Schopenhauer sometimes presents these as an organic unity, he more often dwells on the conflicts between strata and the failure of our allegedly human part (that is, reason) to fully govern our lives. According to the philosopher, music, body, and world constitute a series of redoubling multiplicities, each of which is reflected to some degree in the others. It seems plausible, then, that experiences of music could promote two kinds of expanded awareness: first, of the body's own plurality, and second, of the connections between that plurality and the world at large. As we will see, however, Schopenhauer's privileging of one of these symmetries (music and world) over the other (music and body) means that we must depart from conventional readings of his aesthetic theory to rediscover the bodily element he neglected to articulate.

Polyphonic Ecologies

If Schopenhauer was familiar with contributions to music aesthetics by Romantic contemporaries such as Hoffmann, he must have thought he had finally reconciled music's human provenance with its apparent capacity to speak with the voice(s) of nature. Music, for Schopenhauer, creates the semblance of all worldly grades of existence seeking to become expressive. Imaginary conversations aside, Schopenhauer clearly did not think tenors sounded like trees, nor altos like asps. His analogy between music and world does not hold between different orders of sound production but between the degrees of articulateness and (seeming) self-awareness of musical voices and natural entities, respectively. Just as the voices of the music Schopenhauer knew best get more elaborate as they ascend in register, so too did the world present the philosopher with an apparent ontological ascent from the rocks underfoot to the humans striding atop them. In music, it is as if everything in the world raises its voice in a stratified chorus of being—everything including the "trees, flowers, animals, stones, [and] water" apostrophized in Hoffmann's *Kreisleriana*.[23]

Although Bryan Magee is correct in viewing Schopenhauer's philosophy as essentially modern in spirit, the basic principles of his musical aesthetics have clear precedents in Romantic thought.[24] The famous pronouncement at the end of *Kreisleriana*—"Our kingdom is not of this world, say musicians, for where in nature do we find the prototypes for our art, as painters and sculptors do?"—bears more than a passing resemblance to Schopenhauer's claim that music does not represent things of the world in the manner of the other arts.[25] Yet neither Hoffmann nor Schopenhauer left the matter there; Hoffmann, for his part, went on to praise the intimate relationship between music and the "whole of nature." In a passage of compounding strangeness, he describes the process by which musicians learn to recognize the "secret music of nature": "The audible sounds of nature, the sighing of the wind, the rushing of streams, and so on, are perceived by the musician first as individual chords and then as melodies with harmonic accompaniment." For the skilled composer, music eventually becomes "a universal language of nature" that "speaks to us in magical and mysterious resonances" not conveyable in words.[26] It is as if music and nature gang up against language, luring humans into a realm where voices denote no subjects and speech has no vocabulary. In Hoffmann's essay, these statements come just after a story about a young man for whom the song repeatedly sung by his father and the patterns of moss in his father's garden merge, in the young man's vivid imagination, into a single, multisensory, aesthetic apparition that straddles natural and human domains.

Schopenhauer reinterpreted the "spirit of nature" Hoffmann heard in music as will, a move that, given the philosopher's equation of will and body, has the potential to imbue Romantic aesthetics with a more robust sense of embodiment. Why, then, did Schopenhauer attempt to remove the body from aesthetic experience by describing the latter as pure, will-less contemplation? Not surprisingly, this aspect of the philosopher's theory has occasioned considerable bewilderment among his commentators. Stephan Atzert complains that Schopenhauer left no room for the feelings occasioned by musical listening—feelings which, Atzert argues, account for the art's elevating effects.[27] Friedrich Nietzsche also alluded to the limitations of his predecessor when he made note of "*Schopenhauer's* scandalous misunderstanding when he took art for a bridge to the denial of life."[28] Indeed, statements such as the following, from the first volume of *The World as Will and Representation*, seem to indicate the immobilization of will in musical experience: "The inexpressible depth of all music, by virtue of which it floats past us as a paradise quite familiar and yet eternally remote . . . is due to the fact that it reproduces all the emotions of our innermost being, but entirely without reality and remote from its pain" (*WWR*, 1:264).

Of course, one can imagine recognizing musical attempts to evoke joy or sadness without experiencing those emotions oneself. Yet if music succeeds in transporting us beyond the reach of everyday emotions, it does so only to land us in what can be an intensely somatic realm of aesthetic feeling. As Schopenhauer exclaims: "But how marvelous is the effect of *minor* and *major*! How astonishing that the change of half a tone, the entrance of a minor third instead of a major, at once and inevitably forces on us an anxious and painful feeling, from which we are again delivered just as instantaneously by the major!" (*WWR*, 1:261). That musically motivated feelings of pleasure and pain are more than purely intellectual is evident in Schopenhauer's disagreement with Gottfried Wilhelm Leibniz's designation of music as "an unconscious exercise in arithmetic" (*exercitium arithmeticae occultum*; quoted in *WWR*, 1:256). If music were only that, Schopenhauer argues, then "the satisfaction afforded by it would inevitably be similar to that which we feel when a sum in arithmetic comes out right, and could not be that profound pleasure with which we see the deepest recesses of our nature find expression" (*WWR*, 1:256). In even less ambiguous terms, he asserts in volume 2 of his treatise that music "acts directly on the will, i.e., the feelings, passions, and emotions of the hearer, so that it quickly raises these or even alters them" (*WWR*, 2:448). According to the logic of Schopenhauer's philosophy, music simply could not have what he calls such a "powerful and penetrating" effect without engaging the listener's will—which means engaging the body and its capacity to feel (*WWR*, 1:257).

Even if one follows Schopenhauer, then, in holding that music makes audible only "the innermost soul of the phenomenon without the body" (*WWR*, 1:262), it does not follow that the bodies of listeners are merely passive receivers of music's bodiless gesticulations. But although music's movements and affects are not ascribable to those of specific bodies, it is not quite correct to say that they are disembodied. Music's will-like fluctuations not only arise from the actions of performing bodies but also constitute sonic analogies of embodied motion and feeling that become active partners in own our physical and emotional comportment. Music limns a virtual body's possibilities for movement, affect, mood, feeling, or what have you. One can agree with Schopenhauer that music "does not express this or that particular and definite pleasure, this or that affliction" without making the more extravagant claim that music expresses "joy, pain, sorrow, horror, gaiety, merriment, peace of mind *themselves*" (*WWR*, 1:261). In other words, music can transcend the particular without transmitting the essence. This nonmetaphysical transcendence is precisely what accounts for music's broad appeal and its capacity to be experienced in so many ways by individual listeners, who grasp music's expressive import with the interpretive assistance of their own bodies—through unconscious and cross-modal associations between rhythm, register, affect, movement, dynamic intensity, textural density, and so on.

Bringing Schopenhauer's aesthetics back in line with the rest of his remarks on the identity of will and body makes possible an interpretation of his musical ecology that dramatically expands the scope of the embodied experiences listeners might seek out in music. Just as intuition opened the door to Schopenhauer's recognition of will in the bulldog ant, the *Mimosa pudica* (sensitive plant), and magnetic ore, so too did intuition allow him to propose an analogy between music's wordless voices and the myriad channels of will coursing through animals, vegetables, minerals, and humans (*WWR*, 1:116, 147–48). Music, he surmised, "expresses the innermost nature of all life and existence" (*WWR*, 2:406). In short, music offers humans a way to confront their nonhumanness, or rather, their more-than-humanness, including the innumerable material and physiological processes that Schopenhauer unsuccessfully tried to eliminate from the Idea of the human.

Music also allows us to indulge our humanness, if we accept Schopenhauer's description of melody as the voice which most clearly communicates humanity, along with all the cognitive content that implies. But this too turns out to be a slippery proposition. Schopenhauer initially associates melody with the freedom of thought, a perspective perhaps best justified in the case of solo texted music. He calls melody the objectification of "the intellectual life and endeavor of man." A moment later, however, he declares that melody relates the "secret history of the intellectually enlightened will." That history consists of "every agitation, every effort, every movement of the will, everything which the faculty of reason summarizes under the wide and negative concept of feeling, and which cannot be further taken up into the abstractions of reason." At this point, we have moved pretty far away from a notion of melody as the "uninterrupted significant connection of *one* thought from beginning to end, and expressing a whole" (*WWR*, 1:259). Instead, we find ourselves plunged into the disunified realm of feeling, whose turbulence speaks to the supremacy of will rather than reason. This is the "paradise" where music conjures up "all the emotions of our innermost being." No unity, no single thought holds sway here. As Schopenhauer explains, "I know my will not as a whole, not as a unity, not completely according to its nature, but only in its individual acts" (*WWR*, 1:101). This realm of feeling is one that humans share with animals, for the will, the philosopher alleges, "is essentially the same in all animals as that which is known to us so intimately. Accordingly, the animal has all the emotions of humans, such as joy, grief, fear, anger, love, hatred, strong desire, envy, and so on."[29]

Melody, then, puts us in touch with human thought as well as the "animal" realm of feeling. But why stop there? Music's sonic excitations travel along multiple perceptual, physiological, affective, and cognitive pathways.[30] Music engages our motor systems and spatial imagination, creates virtual scenes in which we participate, and coaxes responses from us ranging from the highly refined to the inscrutable and involuntary.[31] Music brings evolutionarily ancient physiological systems into contact with more recent cognitive developments, and in this respect the locus of its impact messily exceeds the rational center of the Idea of the human. Music ramps us up or calms us down; it couples in little-understood ways with our sympathetic and parasympathetic nervous systems.[32] We sometimes respond to music in the manner of a reflex—or, as Schopenhauer might have put it, as a plant responds to stimuli—for example, by involuntarily shedding tears or getting chills.[33] Music intervenes in the "vegetative" functions of breathing or heart rate, allowing it to be used therapeutically or in service of meditation. Finally, music acts on the sheer physical materiality of our bodies, as our flesh moves in sympathetic vibration with the tones of the bass. Music may not be a purely physical force, but it often acts like one: we speak of being swept, blown, or carried away by powerful pieces of music, just as if they were gales or hurricanes.

The diversity of responses music is capable of eliciting and the multiplicity of responses in play during any given moment of listening testify to our physiological and psychological complexity or, in Schopenhauer's terms, to the multiple grades of will that permeate our bodies. Music activates nested layers of our material instantiation, from the physical reactivity of flesh to vibration and reflex-like reactions to auditory stimuli to the excitation of "animal" feelings and reflective musings over form and meaning. Is it any wonder that Schopenhauer thought music gave voice to the entire span of will as it wends its way from the mass of the planet all the way to human thought? If only he could have admitted that music not only echoes the world's ontological chorus but also sets that chorus singing within the bodies who play and listen.

Readers suspicious of the concept of will might find it more congenial to think about Schopenhauer's musical ecology in terms of energy. Music transforms sounds into energetic patterns whose formal properties have many analogues in the natural world. For instance, music proliferates by way of self-referential processes such as repetition, variation, mutation, expansion, and symmetrical unfolding, patterns found not only in human activities but also in organic growth and the exertion of inorganic forces. In her study of Schopenhauer, Hannan maintains that "our emotions are examples of the same forces, the same buildups and releases of energy, that occur throughout nature, both in the parts we think of as living and the parts we think of as nonliving. Any music that may be interpreted in terms of human feelings, then, could just as well be interpreted in terms of animals or plants, or physical/geological features such as rocks, rivers, and volcanoes."[34] Summing up the everything and nothing that comprises music, Hannan writes, "music manages to be about all natural forms and forces at once, and none in particular."[35] The flip side of music's resistance to stable meanings, in other words, is its plenitude of interpretive possibilities. By flooding our bodies with energies and resonances that saturate the world at large, music allows us to transcend not the body per se but the body's human(ist) limits.

Epilogue on Transcendence

Schopenhauer's aesthetics of music, I have argued, locates us squarely within the complexities of earthly embodiment even as it depicts music as the playground of entirely bodiless feelings. Some readers may find this conclusion rather perverse, in that it flies in the face of everything taken for granted about Schopenhauer's philosophy. To be sure, Schopenhauer considered the "pure subject of knowing," and likewise the pure subject of aesthetic experience, to be completely disengaged from will. But such a stance is, strictly speaking, impossible for anyone still alive. In fact, Schopenhauer's treatise makes it clear that if you wish to engage in pure knowing, you must make certain bodily preparations. He prescribes "a peaceful night's sleep, a cold bath, and everything that furnishes brain-activity with an unforced ascendancy by a calming down of the blood circulation and of the passionate nature" (*WWR*, 2:368).[36] In other words, pure knowing, if such a thing existed, would also be a kind of physiological state, one predicated on the complete satisfaction of bodily needs rather than their transcendence. Here it is worth recalling Magee's observation that Schopenhauer's actual approach to life was diametrically opposed to the denial of the will he recommended in his writings. Magee points for confirmation to the philosopher's "huge range of interests and enjoyments" as well as the strength of his appetites, whether for food, sex, conversation, theater, or music.[37] Perhaps it is only when philosophers have finished eating, making love, and being entertained that they are ready to enter a state in which "consciousness of other things [is] raised to so high a potential that consciousness of our own selves vanishes" (*WWR*, 2:368). What replaces self-consciousness in moments of "pure knowing," Schopenhauer claims, is the rapt contemplation of Ideas, or the purportedly eternal forms of nature's manifestation.

Although I do not fully share Schopenhauer's faith in pure knowing, let alone his hope of "'transcending' the constraints of material life," as Amitav Ghosh describes one of the signature themes of modern thought, I think there still may be something of value in the idea of "vanishing" through aesthetic or contemplative experiences.[38] Aside from the temporary relief such experiences offer from human self-obsession, it is also worth pondering the biological peculiarity, and maybe even the ecological promise, of being a creature who delights in the suspension of some of its most distinctive traits. For it is not just the will that is quieted in instances of pure knowing; reason, according to Schopenhauer, also yields to what is a predominantly perceptual means of engaging with the world. Knowing, for this philosopher, does not consist exclusively, or even primarily, in the operations of conceptual thought. "The kernel of all knowledge," Schopenhauer argues, "is *perceptive or intuitive* apprehension," meaning the discernment of the true "nature of things," or the Ideas that all existing things instantiate (*WWR*, 2:80).[39] To reach this region of truth, we must combat both the desires of the will and the distractions of reason with a practice of "self-denial" (*WWR*, 2:367). We must, in a sense, become another kind of animal, one whose perceptual focus shifts from the particular to the universal. "Animals," Schopenhauer remarks, "have understanding without the faculty of reason, and consequently they have knowledge of *perception*, but no abstract knowledge" (*WWR*, 2:59). For humans, by contrast, "all that is merely imaginable or conceivable, and consequently also what is false, impossible, absurd, and senseless, enters into abstract concepts, into thoughts, ideas, and words" (*WWR*, 2:69).[40] The Schopenhauerian savant is something else again, a *disinterested animal* who ceases to think in order to perceive the truth. It is not only self-consciousness that vanishes in this undertaking but also a conception of the human defined by its powers of reasoning.[41]

Those seeking to trade the fictions of conceptual thought for the reality of Ideas can turn to either art or nature. Art, says Schopenhauer, is a human undertaking that explicitly spurns the meddling of concepts; music traffics in universal movements of the will, while the other arts open a window onto universal Ideas. Yet nature, precisely because it is not a human undertaking, affords even better opportunities to substitute the chatter of mental rumination with the discernment of perception. Art may aspire to represent with human means things that exist independently of the human, but nature achieves this independence by nature. Careful observation of the natural world, Schopenhauer suggests, thus offers temporary respite from the "defects" and "inaccuracies" of human thought. Moreover, he singles out aesthetic appreciation of natural vistas as the *only* one of our many "complicated brain-phenomena" that is "faultless" and "perfect." "A beautiful view," he states, "is therefore a cathartic of the mind, just as music is of one's feelings, according to Aristotle" (*WWR*, 2:403–4). Music purges our emotions of too much reality, while nature purges our thoughts of too little reality. Each recalibrates our humanness along one of the scales that, for this philosopher, measure experience: those of will and representation.

At certain points in Schopenhauer's writings, natural beauty seems to gain the upper hand over artistic beauty as a vehicle for transcending the here and now of individual existence as well as the human entrapment in thought. In volume 1 of his treatise, he maintained that humans possess the greatest beauty because their Idea corresponds to a "high grade of the will's objectivity" (*WWR*, 1:210). But this self-congratulatory move is hard to square with his later assessment in volume 2, which casts the disinterested gaze of the aesthetic on the human species and discovers a Pandora's box of ugliness from which hope is missing. Humanity's exploits generate a litany of woes including "universal need, restless exertion, constant pressure, endless strife, forced activity, with extreme exertion of all bodily and mental powers," all of which serves no purpose other than propagation of the species. Humans self-servingly strive for the common good, only to find themselves in situations where the "blood of the great multitude must flow" (*WWR*, 2:357). Schopenhauer's distaste for the "plotting and planning" that fills human history is perhaps what later led him to correlate beauty not only with timeless Ideas but also, more pointedly, with the absence of human aims altogether. In his "Isolated Remarks on Natural Beauty," he rhapsodizes, "Yet how aesthetic nature is! Every little spot entirely uncultivated and wild, in other words, left free to nature herself, however small it may be, if only man's paws [*Tatze*] leave it alone, is at once decorated by her in the most tasteful manner, is draped with plants, flowers, and shrubs, whose easy unforced manner, natural grace, and delightful grouping testify that they have not grown up under the rod of correction of the great egoist, but that nature has here been freely active. Every neglected little place at once becomes beautiful" (*WWR*, 2:404).[42] Only when nature remains free from the intrusive imposition of human plans, as in uncultivated places or, more cunningly, in the "co-productivity" of English gardens, does it succeed in being "perfectly beautiful" (*WWR*, 2:404; note the traces of Kant's aversion to "stiff regularity" in landscape gardening).[43] Beauty, Schopenhauer contends, is just that property of things which emerges when they are viewed "purely objectively and outside all relation" (*WWR*, 1: 210)—namely, as representatives of Ideas rather than objects of human interference. The beauty of the plant kingdom, where the "objectification of the will-to-live that is still without knowledge" reigns supreme, is exemplary, because the forms of vegetal appearance in the wild are "not determined, as in the animal world, by external aims and ends, but only immediately by soil, climate, and a mysterious third something, by virtue of which so many plants that have sprung originally from the same soil and climate nevertheless show such varied forms and characters" (*WWR*, 1:210).

It is tempting to interpret Schopenhauer's "mysterious third something" as a placeholder for evolution itself, for the adaptations to the living and nonliving environment that make organisms what they are. The experience of beauty may be predicated, for Schopenhauer, on a nonrelational relation between subject and object (or, more precisely, between the "pure subject of knowing" and Idea), but relations within nonhuman contexts, relations among the denizens of ecosystems, turn out to be a crucial factor in the origins of beauty. Flowers evolved into colorful and shapely forms for their pollinators, not for us. That humans have taken no part in such relations yet still find beauty in their results is not only a source of great wonder, but also, I believe, one of the more redeeming characteristics of our species.[44]

In this regard, Schopenhauer's philosophy presents a salutary alternative to what Ghosh identifies as the modernist imperative by which "human consciousness, agency, and identity came to be placed at the center of every kind of aesthetic enterprise."[45] The philosopher's elucidation of a kind of aesthetic engagement that is precisely *not* about us, that turns on a certain *vanishing* of the human, holds out the promise of radically noninvasive and noninstrumental modes of interaction with nature. Schopenhauer identifies the self-forgetting in moments of contemplation as a powerful counterforce to more benighted forms of human intervention in the natural world—a maneuver that is not without contemporary appeal. This is not to say that the global ecological situation would necessarily improve if we all just stopped thinking and contemplated beauty whenever we ran across it. As Gernot Böhme has suggested, the future potency of the aesthetic in "transforming the human-nature relationship" turns not just on contemplation but also on "co-experience or commiseration" with nature through the medium of the body.[46] At the same time, Schopenhauer's philosophy indicates that alongside environmental activism, efforts to develop sustainable technologies, and other thought-intensive ways of furthering ecological conservation, space should be reserved for the perceptually rich thoughtlessness of aesthetic wonder, and for the humility that ideally follows in its wake.

Even after making ample allowances for its many paradoxical aspects, however, Schopenhauer's aesthetics still might be essentially unrealizable today. It is ironic, if consistent with the way philosophical and aesthetic esteem for nature tends to rise when what it apostrophizes comes under threat, that Schopenhauer formulated his antihumanist aesthetics of nature in the midst of a societal transformation that may eventually render those aesthetics obsolete. In the current era of human-generated climate change known as the Anthropocene, the luxury of not mattering has become decidedly scarce. Over the coming decades, the life of every creature on this planet will likely be shaped to some degree by global warming. This is a much darker sort of relationality than the rose-tinted varieties I invoke elsewhere in this book. What is repressed in Schopenhauer's theory of the nonrelationality of beauty—the environmental impact of all those humans traveling around in search of pure knowing—returns with a vengeance. This is a relationality that threatens the very otherness of nonhuman others, by which I mean not their difference from us but their autonomy as organisms. Another way of putting this would be that in the Anthropocene, humans and their by-products become part of every organism's environment.[47] If the prospect of no longer being able to contemplate something that goes its own way despite us, or to encounter inhabitants of the natural world that remain unaffected by our presence, would intensify efforts to protect fragile ecosystems and slow the production of greenhouse gases, then perhaps we all could benefit from becoming a little more Schopenhauerian.

4

The Floral Poetics of Schumann's *Blumenstück*, op. 19

Discussions of Robert Schumann's little-known *Blumenstück* (*Flower Piece*), a short composition for piano published in 1839, typically begin with apologies. Anthony Newcomb notes that in comparison to works like *Kreisleriana* or the Toccata, op. 7, *Blumenstück* is "texturally, technically, and formally a much simpler kind of music," and he concludes that the piece is best considered "a higher level of salon music."[1] John Daverio agrees, observing that *Blumenstück* and its relative, the *Arabeske*, op. 18, "might find a place in the bourgeois salon, or even bring in a tidy profit, but neither could lay claim to being high art."[2] Suspicion of the artistic credentials of *Blumenstück* begins with Schumann himself. In a letter to Henriette Voigt, he admitted that the piece, along with the *Arabeske*, didn't add up to much in comparison to the longer and more varied *Humoreske*, op. 20. Playing on the connotations of both titles, he insisted that he was not to blame if their "stems and figures [were] so tender and weak."[3] To another correspondent, Schumann made his excuses by way of the consumers he had in mind, describing both the *Arabeske* and *Blumenstück* as "delicate—for ladies."[4]

Composed during Schumann's brief tenure in Vienna, *Blumenstück* is cast in an accessible style tailored to the growing market for domestic piano music. The first section of the piece immediately establishes this style: simple melodic profiles, regular phrase structures, and a rangy bass line decorated with grace notes nestle within a well-behaved rounded binary form (example 4.1). Occasional Schumannian touches, such as brief melodic forays into the register of the accompaniment (mm. 3–4 and 17–18) and a two-measure extension left hanging on an unresolved dominant seventh chord (mm. 13–14), add interest but do not threaten the congenial atmosphere. Nevertheless, Daverio observes that the piece is not "as unassuming as it first appears," and he goes

EXAMPLE 4.1. Robert Schumann, *Blumenstück*, op. 19, section I

on to describe it as a double theme-and-variations form in which the second theme takes precedence.[5] Similarly, Erika Reiman argues that the piece evinces a "rondo-like" form shot through with the motivic cross-references characteristic of Schumann's 1830s piano music.[6]

These appraisals suggest that commentary on *Blumenstück* is fated to follow a rather predictable script. After the obligatory nod to Schumann's disparagement of the piece, the critic proceeds not to consign it to the heap of nineteenth-century domestic music but to defend it on the grounds of its formal innovations or hidden motivic relationships. Since these qualities ap-

EXAMPLE 4.1. (continued)

pear to cancel out *Blumenstück*'s accessibility and orientation toward women, dismantling its ignominious association with "ladies" becomes a matter of remasculinizing the piece by drawing attention to noteworthy structural features.

I would like to step outside this charmed circle of musicological apologetics by preserving rather than dispensing with Schumann's gendered designation of *Blumenstück*, an approach that casts a more ambiguous light on the piece's motivic and formal techniques. The interpretive and analytical conundrums that ensue are directly related to the mixed connotations of the piece's title. By the mid-nineteenth century, the species of painting known as the flower piece was viewed as a feminized genre of marginal aesthetic significance. Meanwhile, the popular literature dedicated to *Blumensprache*, or the practice of communicating by way of flowers, deemed its subject matter especially appropriate for women and the expression of conventional sentiments. At the same time, however, Romantic authors admired flowers for their unrivaled power of symbolic evocation. "Nothing is rarer than the flower of romanticism," proclaimed Jean Paul in the *Vorschule der Ästhetik* (*School for Aesthetics*), and in Romantic literature, the flower, whether blue or some other alluring shade, sets many a male hero on a path that passes straight through the feminine en route to the metaphysical.[7] Schumann's *Blumenstück*, I argue, engages both Romantic and sentimental discourses of the flower, confounding categories pertaining to gender, social register, and musical form. Although the hermeneutic approaches I employ in this chapter are more historically oriented than those found elsewhere in these pages, the conclusions I draw feed directly into one of the book's larger themes—namely, how tightly natural entities such as flowers are woven into human self-understanding, even in the case of such seemingly species-specific traits as language and artistic creativity.

Flowers between Allegory and Symbol

In the fall of 1838, Schumann traveled to Vienna in hopes of finding a new publisher for the *Neue Zeitschrift für Musik* (*New Journal for Music*) and establishing a base of operations for himself and his fiancée, Clara Wieck. This venture proved unsuccessful, but Robert did manage to compose several works during his six-month stay. In a letter to Clara from January 1839, he wrote, "I have finished variations, but not on a theme: I'm going to call it *Guirlande* [*Garland*], everything is intertwined in such a peculiar way. There's also a rondolette, a little thing, and then I'm going to take the small pieces, of which I have so many, run them prettily together, and call them 'Kleine Blumenstücke' ['Little Flower Pieces'], like one calls pictures. Do you like the name?"[8] Commentators have reached a variety of conclusions regarding which pieces Schumann was referring to in this passage. Daverio surmises that the rondolette may correspond to the *Arabeske* and that *Guirlande* is lost.[9] As for *Blumenstück*, it might have been one of the "small pieces" Schumann mentioned, or perhaps it comprises several of those pieces "run prettily together." As I will show in a moment, though, the sections of *Blumenstück* are so "intertwined" as to raise the question whether opus 19 actually is the missing *Guirlande*, just with titles switched.

Both headings indicate that Schumann had botanical themes on his mind while in Vienna, but each has its own set of associations. Since antiquity, garlands of flowers have been employed in a variety of ceremonies and rituals. Schumann's musical garland, on the other hand, was likely inspired by thoughts of love. In late eighteenth-century France, aristocratic men routinely presented their beloveds with literary "garlands," or books containing floral images and poems.[10] The trend was inspired by the Marquis de Montausier's *La guirlande de Julie* (1641), a manuscript he commissioned as a birthday present for his fiancée. Once printed, the manuscript appeared in numerous editions throughout the eighteenth and nineteenth centuries. Publishers followed suit with a diverse array of flower almanacs, such as Charles Malo's *Guirlande de Flore*, issued in bindings of assorted colors during the 1810s. Gifts of this sort became common among the middle classes in France after the Napoleonic wars. Schumann's use of the French spelling (*Guirlande*, not

Girlande) suggests that he may have been familiar with the genre—his father was a bookseller, after all—and that his own "garland" was meant as a musical gift for the distant Clara.

The title *Blumenstück*, by contrast, calls to mind a genre of "pictures," as Schumann stated in his letter to Clara. Reiman argues that Schumann's reference to flower painting is an allusion to an allusion—namely, to Jean Paul's use of the term in his novel *Siebenkäs* (1796), whose full title is *Flower, Fruit and Thorn Pieces: Or the Married Life, Death, and Wedding of the Advocate of the Poor, Firmian Stanislaus Siebenkäs*. The actual heading "Flower Piece" is reserved for two digressions midway through the novel, miniature essays on matters of death and religious belief that invoke flower painting as an allegory for the transience of all things earthly. Reiman uses the intertextual link to *Siebenkäs* to support her interpretation of Schumann's piece as embodying both a digressive ethos and the "perfect happiness within limits" Jean Paul attributed to the idyll, a genre he associated mainly with novels written in a "Dutch" (*niederländisch*) or "low" mode.[11] I would suggest, however, that the aptness of the term *niederländisch* lies not so much in the alleged idyllic content of *Blumenstück* and *Siebenkäs* as in the genre of painting that serves as their shared point of reference.[12]

As a branch of still life, flower painting originated in sixteenth-century Europe, but it reached the peak of its popularity and respectability in seventeenth-century Holland.[13] Like other types of still life and genre paintings of everyday scenes, flower painting appealed to the down-to-earth tastes of middle-class Dutch consumers, and its practitioners received some of the highest fees of all artists. The genre's focus on naturalistic detail and its frequent overlap with science went hand in hand with the antihumanism of Dutch Calvinism, as Whitney Chadwick has observed.[14] Women such as Rachel Ruysch made some of the greatest contributions to the genre (figure 4.1); other painters, such as Maria Sibylla Merian (figure 4.2) and Judith Leyster, excelled at botanical illustration, a scientific practice that required formidable attention to detail but was not considered a fine art.[15] The artistic reputation of such illustrations was also compromised by their occasional commercial ties, as in the case of Leyster's watercolors for sales catalogues issued during the Dutch tulip craze of the 1630s.

The seemingly mundane concerns of still life, landscapes, and flower paintings drew scorn from painters engaged in more idealistic pursuits. An oft-quoted critique of northern art attributed to Michelangelo complained that it bestowed too much attention on "external exactness" at the expense of the implicitly internal qualities of reason, symmetry, proportion, and substance. The critique concluded that such art will "appeal to women . . . and

FIGURE 4.1. Rachel Ruysch (1664–1750), *Still Life with Flowers on a Marble Tabletop* (1716). Courtesy of Rijksmuseum, Amsterdam.

also to monks and nuns and to certain noblemen who have no sense of true harmony."[16] Svetlana Alpers has argued that the overriding concern of Dutch painting with the accurate representation of the world upset the economy of the masculine gaze and its pretensions to mastery by placing the world in a more privileged position than the viewer.[17] Alpers proposes that Dutch art demands a mode of perception that attends to qualities such as skill, accuracy, and representational function rather than hidden meanings, symmetries, or

FIGURE 4.2. Maria Sibylla Merian (1647–1717), Solanum mammosum, from *Metamorphosis insectorum Surinamensium* (1705).

proportions.[18] Even the common practice of interpreting flower paintings as allegories, she suggests, passes too quickly over their celebration of the fine details and textures of all-too-perishable materials, in effect downplaying the viewer's imaginary encounter with worldly objects in favor of the contemplation of religious or philosophical meanings.

This trajectory reproduces the reception history of flower painting itself. In the eighteenth century, Dutch women artists helped spread taste for the genre throughout Europe, serving as court painters in Germany and being elected to the Académie Royale in Paris.[19] But tolerance of women artists waned as new ideologies of "natural" femininity circumscribed the professional opportunities available to women. Political philosophers such as Jean-Jacques Rousseau and Denis Diderot preached the restriction of women's activities to the domestic sphere, a call that reverberated in the many late eighteenth-century paintings of mothers with their children.[20] Ultimately, the heroic, masculine themes of Jacques-Louis David's paintings secured the premiere status of historical painting and the exaggeration of sexual difference in nineteenth-century art, to the detriment of women artists and the genres associated with them. Even flower painting's capacity to convey allegorical meanings would not necessarily have impressed critics such as Jean Paul, who, notwithstanding his reference to the genre of still life in *Siebenkäs*, called literary allegories "easy."[21] Similarly, G. W. F. Hegel complained in his *Aesthetics* that allegory is "bleak," the handmaiden of "prosaic universality."[22] Despite the continued practice of flower painting by both men and women, by 1860 one art critic could write, "Let men busy themselves with all that has to do with great art. Let women occupy themselves with those types of art they have always preferred, such as pastels, portraits, or miniatures. Or the painting of flowers, those prodigies of grace and freshness which alone can compete with the grace and freshness of women themselves."[23]

Schumann's designation of his *Blumenstück* as "delicate—for ladies" represents a similarly casual conflation of women and flowers and deprecation of flower painting as a woman's genre. The negative associations of flowers can be detected elsewhere in the composer's writings. In a letter to Clara from April 1838, he complained about the compositions he regularly encountered in Leipzig: "Apart from all their faults of construction, they deal in musical sentiment of the lowest order, and in commonplace lyrical effusions. The best of what is done here does not equal my earliest musical efforts. Theirs may be a flower, but mine is a poem, so much more spiritual; theirs is a crude natural impulse, mine the result of poetical consciousness."[24] Schumann's words imply that musical "flowers" say only the most banal of things, that no real expertise is required to create them, and that their significance is ephemeral, fated to wilt as quickly as the real thing. Such compositions, it seems, failed to join nature and spirit in an ideal Romantic union.

Daverio's suggestion that "we would expect a 'flower-piece' to be graceful and elegant, but little more" seems at first blush to accord with contemporary views.[25] How, then, to explain Schumann's remarkable musical homage to flowers in his setting of Heinrich Heine's "Am leuchtenden Sommermorgen" ("On a Bright Summer Morning"), the twelfth song of *Dichterliebe* (*A Poet's Love*)? Wandering around in his garden one summer morning, the lovelorn poet hears the flowers speaking among themselves. To capture this impossible sound, Schumann opens an enharmonic gateway to another world: on the words "sprechen die Blumen" ("the flowers speak"), the augmented sixth chord which began the song assumes the guise of a (respelled) dominant seventh on F-sharp, steering the music toward what amounts to an apparition of B major within the tonic environment of B-flat (example 4.2, mm. 8–9). While the vocal line approaches the right place from the wrong direction, so to speak, with a convincing descent to C-flat on the word "Blumen," the

EXAMPLE 4.2. Schumann, *Dichterliebe*, "Am leuchtenden Sommermorgen," mm. 1–11

piano's right-hand arpeggios touch on dissonances (C♯ and A) that push the harmony past B major (to a B dominant seventh) before actually having arrived there. The spectral voices of the flowers remain just beyond the poet's grasp; he remains mute (*stumm*) in their presence as the music returns to the home key of B-flat (m. 11). Moments later, the poet begins to understand. Adopting a slower and quieter tone, the flowers relay a secret message from the parallel universe of G before reverting to the relative minor g (mm. 17–20, not shown) on the word "trauriger" (sad). At home in the land of fairy tales and endowed with a miraculous faculty of speech, these flowers clearly have more to say than the ones whose "lyrical effusions" Schumann scorned.

Dichterliebe as a whole illustrates Schumann's (and Heine's) fascination with floral imagery; eight of the sixteen poems bestow some degree of symbolic significance on flowers and other plants.[26] Moreover, Schumann himself marshaled the trope of flower speech in moments of heightened critical fervor. In a review published in the *Neue Zeitschrift* in 1838, he lavished praise on a string quartet by the composer Hermann Hirschbach: "Words seek in vain to describe how his music is formed, what all it depicts: his music is itself speech, such as the flowers speak to us, as eyes narrate mysterious tales, as kindred spirits communicate with one another over vast lands; the speech of the soul, the most inward, rich, and genuine musical life."[27] In this instance, flowers trespass their sentimental boundaries to speak poetically indeed—to speak up when description fails, to speak metaphorically.[28] Schumann had ample precedent for such flights of fancy: playing on the trope of florid rhetoric, Jean Paul called metaphor a "small poetic flower" that joins body and spirit in a single image.[29] For Schumann, the flower not only figures figuration but also serves as a symbol in its own right. Like music, flowers seem to speak the "inward" language of the soul, and those who can decipher their speech—or think they can—accede to the inner circle of Romantic enthusiasts. Where, then, does the flower find its native ground? In the realm of "commonplace" sentiments and "easy" allegories, or among those devices most esteemed in nineteenth-century aesthetics, symbols?

Saying It with Flowers

Schumann's musical and critical allusions to *Blumen* point to the multiple expressive registers that flowers occupied in nineteenth-century culture. Although flowers had long been associated with "convention and domesticity," in Beverly Seaton's words, Romantic authors charged them with expressing "higher thoughts" and "nobler feelings."[30] Novalis famously captured the lofty connotations of the flower in his novel *Heinrich von Ofterdingen* (1800). On the first page, the aspiring poet Heinrich muses, "I yearn to get a glimpse of the blue flower. It is perpetually in my mind, and I can write or think of nothing else . . . such a strange passion for a flower is something I never heard of before."[31] Friedrich Kittler has drawn attention to the curious metaphysics at work in such narratives. According to Kittler, the concepts Nature, Love, and Woman were synonymous in the so-called discourse network of 1800, because they all stood for the idealized origin of poetic inspiration.[32] Kittler posits that new phonetic methods of reading, typically intended for mothers to use in teaching their children, advertised a more natural relationship between speech and reading than did letter-bound memorization. The mother, in her role as the avatar of Woman, transmitted these methods to her children, an arrangement that made her the agent who calls forth the reader's inner voice. Kittler argues that the Romantic trope of the inner voice as the origin of autonomous creativity represents the pedagogical voice made transcendental.[33] Nature, Woman, and Mother mutely inhabit this origin, inspiring love, and with it poetry, in their male devotees.

For Novalis's Heinrich, poetry begins with a blue flower. Staring at the flower in his longed-for dream, he sees a beautiful face inside it. Just as the flower seems about to speak, his mother's voice wakes him from slumber.[34] The distance between dreams and the waking world precisely measures the distance between the transcendent source of poetry, here rendered as the triad Flower/Woman/Love, and its empirical corollary, Heinrich's actual mother. The budding poet's masculine inwardness flourishes in the abyss that separates Woman from women.

In other Romantic tales, as in Heine's "Am leuchtenden Sommermorgen," the inner voice is heard as the speech or singing of flowers. E. T. A. Hoffmann's "Ritter Gluck" (1814), for example, tells of a young man's encounters with a mysterious old composer who entrances him with cryptic stories about the "kingdom of dreams" where true artistry originates. Once, the composer relates, he awoke from this dream into another, in which he witnessed "a large, bright eye that was looking into an organ; and as it looked, tones sounded forth and shimmered and entwined themselves in marvelous chords that had never before been conceived."[35] The eye promises to share its melodic prowess with the composer, a promise fulfilled when the masculine eye joins forces with the feminine flower:

> For many years I sighed in the kingdom of dreams—there—indeed, there—I sat in a marvelous valley and listened to the flowers singing together. Only a sunflower was silent and sadly bowed her closed calyx to the ground. Invis-

ible bonds drew me to her—she raised her head—the calyx opened and shone toward me from within the eye. Now tones, like rays of light, flowed from my head to the flowers, which greedily drank them. The leaves of the sunflower grew bigger and bigger—fire streamed from them—encompassed me—the eye had vanished and I was in the calyx."[36]

Tapping into the transcendent source of creativity represented by the sunflower, the composer proceeds to embark on his greatest works. That era, however, has long passed: at the end of the tale, the composer cries out for the day when "the sunflower should raise me again to the eternal."[37] One can well imagine some variation of this line as the epigraph to Schopenhauer's *The World as Will and Representation*, the first volume of which was published five years after Hoffmann's story.

The singing of flowers migrates from dreams to the supernatural realm in the "whispering" and "lisping" that Anselmus, the main character of Hoffmann's "The Golden Pot" (also 1814), hears coming from the blossoms of an elder tree, a sound that Kittler calls "an auditory hallucination of the Mother's Mouth."[38] At first, Anselmus does not understand the natural language of blossoms, breezes, and sunlight swelling around him on Ascension Day. But as soon as the spectral voices congeal into the figure of a green snake with deep blue eyes—Goethe's *Märchen* meets Novalis's blue flower?—Anselmus falls in love, the first step in learning to decipher nature's speech. Significantly, this love does not take a normal human woman as its object but rather the otherworldly creature Serpentina, daughter of the salamander-turned-archivist Lindhorst.

In cases such as these, flowers speak to men, and men only, from the Nature/Woman/Love nexus at the root of poetic inspiration. Yet, as Kittler points out, Anselmus understands the utterances of nature not instinctively but through the intervention of Lindhorst, a circumstance that points to the careful management of the "language channel" (Kittler's term) by which Nature or Woman speaks to aspiring poets. To be invested with transcendence, the flower had to be dissociated from its usage by actual women and transposed into a mystical realm. Hoffmann's tale accomplishes this by elevating Lindhorst's knowledge of nature over that of the witch who aids Veronica, Anselmus's real-world love interest. At once father figure and mythical being, Lindhorst commissions Anselmus to copy Arabic manuscripts and other exotic texts, in effect training him to read the book of nature.

That Hoffmann associated recondite knowledge of nature with Eastern cultures while isolating women from such knowledge testifies to the complicated heritage of the Romantic trope of flower speech. Although symbolic meanings have been attributed to flowers since ancient times, the idea that flowers speak a decipherable language has its origins in the reports of visitors to Istanbul in the early eighteenth century. The best known of these are Aubry De La Motraye's accounts of his travels in Europe, Asia, and Africa (1723) and Lady Mary Wortley Montagu's *Turkish Embassy Letters*, based on visits in 1717 and 1718 but not published until 1763.[39] Both authors describe a method of communication practiced by Turkish lovers that made use of flowers, fruits, feathers, and other natural objects. Each object was associated with a specific verse, so that, as Montagu asserted, one might "quarrel, reproach or send letters of passion, friendship or civility" with objects alone.[40] Both Montagu and Motraye called the examples they unearthed Turkish love letters, and Europeans who read their reports assumed that the technique allowed private messages to pass between women and men.[41] The language came to be known as the *salām* (Arabic for "peace," also used as a greeting).

With its tantalizing implications of forbidden love and secret codes, the *salām* galvanized the imaginations of Romantics flush with the desire to penetrate nature's hidden meanings. Seaton's study of flower language illustrates how Westerners viewed the reports from Istanbul through their own interpretive lenses, choosing flowers and plants from the wealth of objects included in the *salām* and projecting symbolic meanings onto their uses in this exotic Turkish practice. Goethe and Heine both made mention of the *salām*, as did Novalis in *Heinrich von Ofterdingen*.[42] Toward the end of the novel, the young hero falls into conversation with a physician named Sylvester. The physician praises "living nature," deeming plants "the most immediate language of the soil." Reflecting on the beauty of flowers, he marvels at the earth's "green mysterious carpet" whose "strange script is legible only to its lover, like the floral bouquet of the orient. He will read forever and yet never be sated, and he will daily become aware of new meanings, new more ravishing revelations of loving nature."[43] Like others at the time, Novalis assumed that Turkish bouquets communicated symbolically—that is, in the open-ended, polysemic manner prized by Romantics. Novalis's Westernized version of flower language replaced the one-to-one correspondences documented by Montagu and Motraye with a symbolic economy in which "new meanings" are always possible.

This conception of the *salām* was widespread enough that the well-known Austrian orientalist Joseph Hammer (later Hammer-Purgstall) felt compelled to debunk it in an 1809 essay entitled "Sur le langage des fleurs" ("On the Language of Flowers").[44] First of all, Hammer insisted that the *salām* originated in the royal harem of Istanbul and was not in general use. Second, he explained that the links between objects and phrases of the *salām* did not consist of resemblances or associations (in the manner of metaphor or metonymy), but of assonances. That is, the *salām*'s inventors devised phrases that rhymed with the names of common objects and then used those objects as shorthands for the phrases. Choosing one of Montagu's examples, Hammer noted that the name for "pear," *armoude*, rhymes with "hope," *omoude*; pears then come to stand for the phrase "give me some hope." His essay enumerates some two hundred entries in the *salām*'s vocabulary, which he gleaned from interviews with Greek and Armenian women connected to the royal harem.

Because of its conventional basis, Hammer did not believe that the *salām* was used for communication beyond the harem walls. He argued that if the phrases associated with flowers and other objects were commonly understood, as accounts of the practice implied, then using them to compose secret missives would have been far too risky. Instead, he proposed that women living in the harem invented the language as either an amusement or a "cipher for lesbian declarations."[45] Hammer's claims thus turned the Romantic notion of flower speech on its head. In the *salām*, natural objects "spoke" a language, all right, but their meanings were the self-conscious product of human invention rather than a fantastical "secret script" (pace Novalis) issuing directly from nature. More radically, Hammer's contention that the *salām* may have been intended for use between women, speculative as it was, disturbed the Romantic image of flower speech as a medium joining a transcendent Nature or Woman to man, as in *Heinrich von Ofterdingen* or "The Golden Pot."

The explosion of literature devoted to flower language in the early nineteenth century preserved the reliable economy of signification outlined in Hammer's account of the *salām* while suppressing the human invention involved in deciding what exactly flowers signified. Like the literary *guirlandes* exchanged among couples, any number of manuals and pocket books claiming to reveal the secret meanings of flowers and other plants could be found in middle-class drawing rooms of the early and middle 1800s. As the *Biedermeier* counterpart to Romanticism's more rarefied floral symbolism, such guides instructed prospective lovers in the art of sending messages with a well-chosen nosegay, bouquet, or stem. Authors drew on a wide variety of cultural sources (ancient and medieval in addition to Eastern) in compiling lists of meanings, but most of them made it seem as though such meanings could be divined from the plants themselves—a feature that places this literature only a short step away from its Romantic cousin. A flower might say anything from a single concept (e.g., an olive branch for peace) to full-fledged statements, questions, or pleas. The anonymous handbook *Allgemeine Blumensprache nach der neuesten Deutung* (*General Language of Flowers according to the Latest Interpretation*; 1837), for instance, consists of a thirty-page list of plants with tender phrases attached. Hand someone a bunch of violets, and you will be saying, "Our love shall blossom in quiet secrecy" ("In stiller Verborgenheit soll uns're Liebe blühen"). Hand over a sprig of wintergreen, and you will affirm, "I remain constant" ("Ich bleibe standhaft").[46] Schumann's choice of the title *Myrthen* for his gift of songs for Clara (op. 25) indicates his knowledge of this discourse: the myrtle's profusion of blossoms and affiliation with the goddess Venus connects it with love.[47]

More often than not, publications dealing with the language of flowers were expressly geared toward women, as titles like *Neueste Blumensprache... Eine Frühlingsgabe für das schöne Geschlecht* (*The Most Recent Language of Flowers... A Springtime Gift for the Fair Sex*; 1825, by Karl Wilhelm Ewaldt) attest.[48] "What captivates the heart full of feeling," asks the anonymously authored *Neue vollständige Blumensprache* (*New Complete Language of Flowers*; in its eighth edition by 1850), "especially the heart of a delicate, flourishing maiden, more than Flora's lovely children?"[49] Perhaps the most influential of early flower guides, Charlotte de Latour's *Le langage des fleurs* (1819), was translated into German and published in 1820 as *Die Blumensprache, oder Symbolik des Pflanzenreichs*.[50] Latour's text begins, "Happy is the maiden who does not yet know the foolish joys of the world and to whom plant lore affords a sweet occupation ... A garden is for her an inexhaustible source of instruction and elevation."[51] Dispensing with the theme of arcane knowledge, Latour reassured her readers that mastering the language of flowers presents no special challenges. "No deep research into this language is required," she wrote; "nature herself is the best guide. One need learn only a few rules, which shall be set out here, and consult the list of each flower's meanings, to understand this language."[52] The dictionary of meanings at the end of Latour's text is arranged by affect, so that the reader need only look up the feeling or idea she wishes to express to find the appropriate flower or plant: daisies for innocence, roses for beauty, and so on.

In sum, the flower language marketed to women adopted a tone quite different from that of the speech overheard by sensitive souls such as Heinrich von Ofterdingen or Anselmus of "The Golden Pot." In Romantic literature and criticism, the flower as symbol points to the inexhaustibility of nature, especially the limitless creative inspiration that flows from intimate contact with the natural world. Despite Latour's Romantic-sounding appeal to "inexhaustible" gardens, *Blumensprache* primers generally attempted to translate the meanings of flowers into pithy verbal equivalents. Such acts of translation violated the value system of Romanticism by converting symbol into sign, metaphor into allegory. Instead of reflecting in wonderment on the "new meanings" constantly emerging from flowers (as Novalis's physician recommends), those intent on unlocking nature's secrets now had only to consult a dictionary. The *Blumensprache* literature, in other words, turned gifts of flowers into "easy" allegories, to recall Jean Paul's damning appraisal. If flowers themselves now yielded to interpretation so painlessly, it is no wonder that nineteenth-century critics had so little regard for *paintings* of flowers. Perhaps Schumann was justified after all in distinguishing his own musical "poems" from musical "flowers," those vehicles of "commonplace" sentiment that seemed to say just one thing.

As Schumann's own *Blumenstück* reminds us, however, such distinctions of register were not always so clear. For instance, Johann Daniel Symanski's *Selam, oder die Sprache der Blumen* (*Selam, or the Speech of Flowers*), an exhaustive survey of ancient and contemporary flower imagery, casts itself as a scholarly account of flower language for "gebildete," or educated readers. Replete with footnotes and a hundred-page *Blumenkorb* (flower basket) of German poetry on the theme of flowers, Symanski evidently did not intend his book for the readers Latour wooed with assurances of easy learning. Nor did Symanski shy away from Hammer's risqué insinuations, as did Latour: his *Selam* reproduces the orientalist's account of the origins of flower language almost verbatim, up to and including the "lesbischer Vergnügungen" (lesbian pleasures) allegedly enjoyed by Turkish women confined to the harem.[53] But although the *Blumenkorb* includes such venerable examples as Goethe's poem "Die Metamorphose der Pflanzen" (a reference to his earlier treatise) and Schiller's "Die Blumen," the collection also features sentimental allegories including Anton Platner's "Blumensprache"—an 1805 poem later set to music by Franz Schubert—and J. M. Miller's "Blumen und Mädchen" ("Flowers and Maidens"), with its all-too-obvious warnings to maidens not to let flatterers pluck the flower of their innocence.[54] Symanski also provides a dictionary, as did other purveyors of *Blumensprache* handbooks.

The indiscriminate mixture of floral imagery in Symanski's *Selam* suggests another way to understand the relationship between Romantic and sentimental (or *Biedermeier*) discourses of the flower. Kittler argues that male authorship circa 1800 required the complementary function of "feminine reader," whose task is to be edified rather than merely pleased by literature.[55] With their images of transcendently pure maidens "captivated" by flowers (*Neue vollständige Blumensprache*) or "instructed" by their gardens (Latour's *Le langage des fleurs*), flower guides provide a remarkably clear instance of the production of such readers. These handbooks suggest that, to function properly, the Romantic ideology of Woman as the hidden provocateur of art necessitated a certain kind of training—namely, training women to embody Woman. To further this project, sentimental flower guides reconfigured the *salām*—a language apparently invented by women, perhaps even for use among women—as a language for women to use to communicate silently with men. In other words, such guides placed women in the position of Woman, or better, of Woman/Nature/Love, whose gifts of flowers speak "what words dare not express," as Platner's "Blumensprache" affirms. Whether Romantic, *Biedermeier*, or somewhere in between, European conceptions of flower speech can be viewed as attempts to build a "language channel" joining men to Woman, Nature, and Love—the trifold source of creative inspiration.

But, as ever, pristine ideals come with no guarantees. *Blumensprache* guides did not always agree on the supposedly inherent meanings of flowers, and even the protagonist of "Am leuchtenden Sommermorgen" does not appear to benefit from his prodigious intercourse with the natural world. Heine's "pale man" seems not to have been able to transfer his affections from the empirical to the transcendental realm—he fails, that is, to sublimate his love for a woman and devote himself instead to Woman. The channel leading from flower speech to poetry is blocked, and Schumann delineates the poet's silence by answering the flowers' admonishments with a wordless melody in the piano (mm. 23–26). Two songs later, in "Allnächtlich im Traume," the poet sees his beloved in a dream. She speaks a word and hands him a bouquet of cypress—a reference to the *salām*-inspired practice of communicating with flowers and plants, if one that places little faith in the capacity of inanimate things to speak for themselves. When the poet awakens, the bouquet is gone and he has forgotten the word. If only he had remembered: his beloved likely uttered something along the lines of "death," "grief," or "sorrow," all conventional connotations of cypress. Heine's poetry suggests that *Blumensprache* manuals, their optimism notwithstanding, could not ensure the legibility of flowers, nor the legibility of the human beings whose thoughts they were supposed to illuminate.

Origins, Proper and Improper

How might the many-layered trope of the flower inform a hearing of Schumann's musical *Blumenstück*? As Kittler argues, the Romantic ideology of creative production depends on a strict separation between the artwork's origin and its destination: art originates *from Woman* and travels *toward women*. In Schumann's *Blumenstück*, however, origin and destination appear to be reversed: the composer's comments about the piece suggest that it originated not *from Woman* but *from women*—or rather, from the composer's knowledge of actual women and their consumer tastes, even their preference for pleasure over edification. In its delicacy and appeal, *Blumenstück*'s "flower" corresponds to the end, not the origin, of aesthetic production. The flower has become product rather than producer.

As Kittler demonstrates, however, the transcendental origin of creativity is produced as well—produced by the material paraphernalia of pedagogy and cultivated in the hothouse of philosophical Idealism. In a different formulation of the problem of origins, Paul de Man argues that the abundant natural imagery of Romantic poetry attests to a "nostalgia for the natural object, expanding to become nostalgia for the origin of this object."[56] Appeals to the organicism of art often express just this nostalgia for natural origins and natural development. Glossing Friedrich Hölderlin's poem "Bread and Wine," de Man suggests that the poet's call for words that "originate like flowers" speaks to an impossibility we would prefer to forget.[57] As human beings, *we* may originate like flowers, as incarnations of autonomous biological processes, but our own creations are steeped in the contingencies of time, place, and individual choice. Phenomenologically speaking, musical works may seem to enjoy a degree of necessity and autonomy that rivals that of flowers (see chapter 1), but the creative processes that led to those works likely included fits and starts, redactions and revisions.

Schumann's piece highlights the problem of origins not only in its creative motivations but also in its handling of form. Organized as a string of little pieces marked with Roman numerals, the second and fourth of which repeat exactly or as variations, *Blumenstück* combines aspects of rondo and theme-and-variation forms. The technique of motivic "intertwining" ties the sections together through, among other things, frequent melodic recourse to the interval of a fourth. Opus 19 could thus be described, like its shadowy partner *Guirlande*, as "variations, but not on a theme," a concept that rather obviously destabilizes a standard musical principle of origination and nudges the piece into the realm of formal hybrids. As *Blumenstück* unfolds, it gradually becomes clear that the *second* section—or at least its first fourteen measures—constitutes the refrain of this "rondo-like" piece (Reiman's phrase), even though the section is initially presented not in the tonic key of D-flat major but in the dominant, A-flat. The formal origin of the piece is split, divided with respect to tonal and thematic functions. The two come together only at the end, when the section II material is triumphantly stated in D-flat major.[58] Mirroring the reversal of Woman and women, the origin of *Blumenstück* is in a sense found only in its destination.

The relationship between melody and motivicism nonetheless complicates the music's mode of address—that is, its evident targeting of "ladies." Here we must tread carefully as we come face-to-face with the motivic connections that span the piece's sections. Recent studies of Schumann's music have attempted to reconcile local motives with global form, both in single-movement and multimovement works. Peter Kaminsky's influential essay on the composer's piano cycles, for instance, attempts to identify cases in which the "process of cross reference is integral to large-scale tonal and formal structure."[59] It is not hard to discern here the sort of maneuver that Naomi Schor documents in her study of aesthetics and the feminine, a maneuver not unlike that which leads to the devaluation of such genres as flower painting: the subordination of details (or parts) to wholes, and, ultimately, of the (culturally) feminine to the (culturally) masculine.[60] Listening to Schumann's motivic work solely in relation to large-scale form involves an appeal to a mental construct—the whole—that stands at a certain distance from the music as sensuous phenomenon. This is not to deny the obvious truth that some awareness of larger formal processes accompanies the apprehension of musical details as they occur in time but rather to resist the traditional organicist notion that local motivic traits must consistently serve the articulation of a whole understood in terms of tonal structure.

In the case of *Blumenstück*, such an approach would be banal at best. For most of the piece, the sections simply alternate between tonic and dominant keys; the only exceptions are section V, in E-flat minor, and the B-flat minor variation of section II. One might suggest that the opening leaps from A-flat to D-flat in the melody and bass present in microcosm the harmonic oscillation between tonic and dominant. But this rather bland observation serves only to deflect attention from a more interesting dialogue taking place in Schumann's piece, one that unfolds between motive and melody rather than motive and tonal structure.

Over the first few sections of the piece, *Blumenstück*'s motivic play suggests the disintegration of melody rather than its integration into a larger whole. The proliferation of motives begins in the middle part of section I, where a transposed version of the opening melody replicates its characteristic falling fourth by accelerating it to sixteenth notes in measures 10 and 12 (shown earlier in example 4.1). This variation on a detail only enhances the overall accessibility and clarity of the music by generating *more melody*. In section II (marked "Ein wenig langsamer," or "A little more slowly"), however, the music seems to experience some discomfort over its desire to please. The melody preserves the sixteenth-note figure, embedding it between oscillating seconds (example 4.3). On the second beat of measure 24, the melody pauses on B-flat, hesitating briefly as the accompaniment pattern keeps pulsing away; the ensuing phrase completes the thought with the figure A♭–B♭–A♭ in measure 26 to close in A-flat major. In the next two measures, the motive of a fourth traversed by steps returns with a vengeance in contrary motion

EXAMPLE 4.3. Schumann, *Blumenstück*, op. 19, section II

between soprano and bass lines; the motive's melodic origins are somewhat eclipsed by its symmetrical and rather functional setting. The signs of a retreat of melody become explicit in the middle part of this ternary-form section, which begins with the pickup to measure 33. Jumping into the key of E major (an enharmonically respelled flat VI of A-flat), section II's main melody, now set in the tenor register, drops out of the texture altogether on the downbeats of measures 34 and 36. The receding of melody—the sine qua non of accessibility—draws the listener into a negative space beyond the frame, so to speak, perhaps as a reminder that the abyss of male interiority grounding the composer's creative authority is still present after all.

EXAMPLE 4.3. (*continued*)

The next few bars continue the process of the disintegration of melody, a process that dismantles the easy comprehensibility that at first seemed to class *Blumenstück* among parlor pieces. The ascending melodic fourths of measures 27 and 28 are pared down to their last two notes in measures 37 and 38, generating minor seconds that intensify the cadential effect. What results in the top line is the sequence of pitches D–E♭–C–D♭, presented in more compact fashion measure 41's bass line as part of a tiny two-bar coda. In this guise, one may be forgiven for hearing the BACH motive in transposition and inversion.[61] It is worth recalling that Schumann had studied Johann Sebastian Bach's music since 1831 and copied out part of Bach's *Art of the Fugue* in 1837–38, an activity whose repercussions Daverio discerns in the *Kreisleriana*, op. 16 (1838).[62] But since the BACH cipher in opus 19 is heavily disguised, it might best be considered a fortuitous by-product of the circle-of-fifths progression B♭7–E♭–A♭7–D♭ common to measures 37–38 and measure 41. Section III, another ternary form, revels in this newfound contrapuntal intricacy, as the paired half steps D–E♭ and C–D♭ become the inner-voice axis around which chromatic and circle-of-fifths motion revolves (mm. 43–44) before rising to the top of the register in measures 47–48 (example 4.4). This display of contrapuntal élan and its varied repetition in measures 55–62 enclose a four-measure recollection of section II's melody that again exhibits a certain ambivalence about its pleasing character. The top voice gets "stuck" on the interval of a major second—the bass, meanwhile, tries to pull the music back

EXAMPLE 4.4. Schumann, *Blumenstück*, op. 19, section III

EXAMPLE 4.5. Schumann, *Blumenstück*, op. 19, section IV, mm. 71–80

into the orbit of half steps—before acquiescing to the familiar descending-sixteenths figure and cadence at measure 71 (example 4.5).

What is such a dramatic upwelling of elemental counterpoint doing in a piece designed "for ladies"? Even if Schumann were unaware of the allusion to BACH, this venture into advanced contrapuntal territory pushes *Blumenstück* out of the parlor into an indeterminate aesthetic realm, one that seems to harbor certain pedagogical overtones. I would suggest two ways to understand what is going on here. First, by beginning with simple, appealing melodies and then ushering the music's listeners (and players) into the intricacies of counterpoint, Schumann might be showing the "ladies" where all music really comes from. In this case, it comes from the half step, something very much like Kittler's "minimal signified" or expressive sigh, which, in its utter simplicity, might well supply the perfect image of an absolute origin.[63] Significantly, this source is figured as masculine—as the motivic antithesis of melody, even as the muted trace of Bach—for the consumption of women, in effect preserving the male creator's image of authority and access to creative origins.

On the other hand, the fact that we get to the magical appearance of the minimal signified through an idiom associated with domestic parlors and female consumers only highlights the constructed nature of the transcendental origin of creativity—the way it emerges from a nexus of social meaning which it is then presumed to negate. *Blumenstück*'s image of organic development in reverse—perhaps we should call it organic decomposition—is laid to rest after section III, when a truncated version of section II, having dispensed with the disappearing tenor melody and the derivation of the paired half steps, brings the fragmentation via motives to a close and sends the piece on its way. Section IV resumes the project of motivic "intertwining" as soprano and tenor voices leap and descend through fourths, while the bass's drone on A-flat (set a fourth below rather than a fifth above the tenor's tonic notes) creates a pastoral mood (shown earlier in example 4.5). From the upbeat of measure 73 to the downbeat of measure 74, the tenor line exactly reproduces the pitches of section I's initial melodic gesture, a device that accounts for Daverio's interpretation of *Blumenstück* as a double theme-and-variations form. Likewise, the manic section V (labeled "Lebhaft," or lively; see example 4.6) showcases fourths in the top voice (in mm. 97–98, a leap from B-flat to E-flat, followed by a leap to A-flat and descent to E-flat); the section's contrasting phrase resolutely inverts this interval into impetuous leaps of a

EXAMPLE 4.6. Schumann, *Blumenstück*, op. 19, section V, mm. 97–104

EXAMPLE 4.7. Schumann, *Blumenstück*, op. 19, end of final statement of section II and coda

fifth (mm. 101–2). At the very end of the piece, after a triumphant statement of section II material in the tonic, a short coda recalls the elemental pair of half steps, now presented as G–A♭–F–G♭, in the bass (example 4.7, m. 153), as if to suggest that idealized origins have not been forgotten. Indeed, this figure leads to a normative V–I cadence, complete with 3̂–2̂–1̂ motion in the melody (m. 154). But the motive of a fourth has the last word, as a fragment of section II's melody in the bass reworks the cadence as IV–I (m. 155). The motivic detail, in the end, retains the power to shape events over and against the demands of large-scale tonal form.

Rather than assess Schumann's motivic "intertwining" in relation to familiar notions of organicism, one might compare it to the techniques of combination and interweaving used to make garlands and bouquets. These activities remain firmly rooted in empirical and physical realities; proceeding by trial and error and informed by experience, their driving force is the motion of hands winding together fronds and branches or arranging and rearranging flowers to make a harmonious bouquet. Similarly, hands in motion at the piano—Schumann's hands, the hands of women who played the piece for pleasure if not for edification, hands playing the piece today—weave *Blumenstück*'s profusion of motives into a sonic entity as perishable as a garland of roses or a bunch of lilies. Form as combination, form as arrangement: these conceits find their place at the intersection of Romanticism and sentimentality.[64] Like the *Blumensprache* primers tucked away in so many pockets,

FIGURE 4.3. Barbara Regina Dietzsch (1706–83), *Blue Iris (germanica)* (ca. 1740). © Fitzwilliam Museum, Cambridge.

Schumann's *Blumenstück* troubles the boundaries between elite and amateur modes of expression, along with their gendered connotations.

This interpretation suggests that blue flowers might be discovered not just in the fantasies of aspiring male poets (in an imaginary realm of transcendence that shores up time-bound attitudes about gender), or even in the wild (in the human-transcending realm of nature's self-manifestation, as valorized by Schopenhauer), but in places where humans collaborate with nature in the spirit of arts such as horticulture. Places like botanical illustration: Barbara Regina Dietzsch's painting of a German iris (*Iris germanica*), hovering surreally against a dark background, seems to belong to the nocturnal realm of Heinrich von Ofterdingen's dreams (figure 4.3). Places like the *salām*, whose "langue conventionnelle" Hammer accorded its own variety of genius. Or, finally, in venues like domestic books of dried flowers collected for pleasure or, perhaps, for a loved one who is ill.[65] These blue flowers speak to human encounters with a world whose transience and contingency is cause for celebration rather than despair. Even the protagonist of the first "Flower Piece" in Jean Paul's *Siebenkäs*, who dreams that Christ tells him there is no God, realizes that the role of religion is to transfigure the world of the here and now.[66] Schumann's *Blumenstück* may ask us to do the same.

5

Music between Reaction and Response

Materially dependent on the movement of air and the vibration of material, formally aligned with the melodiousness of birds and the periodicity of ocean waves, music stands at the crossroads where human and nonhuman sound making meet. Humans fabricate wind chimes and Aeolian harps for nonhuman players, while musical imitations of birdcalls and rushing streams imaginatively transpose natural sounds into the virtual spaces of composition. As denizens of a world shot through with the irregular regularities of natural rhythms and the fleeting audibility of the harmonic series, humans make music by rationalizing the rhythmic and resonant character of ambient sounds, only to revolt against their rationalizations by periodically attempting to reintroduce the apparent freedom of nature into the constraints of music. The history of music (in the West, at least) could be described as a dialectically evolving arabesque, as an audible intensification and relaxation of the human will to dominate nature.

Yet music also blurs distinctions between humans and nonhumans, as two familiar myths attest. Orpheus was said to make music that inspired humanlike attention in animals, trees, and even stones. The enchanting voices of the Sirens, by contrast, were believed to reduce passing sailors to the level of animals powerless to resist their song. Max Horkheimer and Theodor Adorno famously used Homer's version of the tale to illustrate the dialectic of Enlightenment. Odysseus, thanks to the labor of his subordinates, survives the Sirens' fatal invitation to relinquish the civilized human self (and, along with it, the "barrier between oneself and other life") even as he enjoys the privilege of hearing their song—but only as a "mere object of contemplation," as the "devitalized beauty" of art.[1]

Adorno's own remarks on popular music suggest that the threat of regression posed by music has never been fully neutralized. Recast in terms that sparked disagreement between Jacques Derrida and Jacques Lacan, the myths of Orpheus and the Sirens portray music as calling forth a *response* in creatures thought merely able to *react* and, contrariwise, stripping away the capacity for *response* in humans, leaving nothing but *reaction* in its place.[2] For Lacan, giving a response depends on the presence of the Other as witness to truth, something he presumed to be absent for animals because they lack language (or, more precisely, access to the symbolic order). Responding thus involves what Derrida calls a "second-degree reflexive power" reserved for humans, a power that suffuses such tactics as pretending to pretend.[3] Derrida questions the "purity and indivisibility" of reaction and response in Lacan's thought, a purity which strikes him as untenable given that, among other things, the psychoanalytic hypothesis of the unconscious complicates any claim that human responses can be fully transparent to themselves.[4] The unconscious, posits Derrida, injects "some automaticity of the reaction in every response," no matter how free that response may seem.[5] Indeed, the current interest in preconscious, biologically rooted affects lends further significance to Derrida's investigation of "the reactionality in the response."[6]

Viewing the problem from another angle, recent studies of animals, especially primates, demonstrate that their cognitive abilities go far beyond the passivity implied by the concept of reaction. The apes that primatologist Frans de Waal describes in moving detail in his book *Our Inner Ape*, for example, appear to treat him, as well as fellow apes, as others to whom, in nonlinguistic fashion, questions are posed and responses owed.[7] Donna Haraway affirms that animals can "engage *one another's* gaze *responsively*," an ability that, she argues, demands more searching philosophical responses than Derrida's own.[8]

In challenging Lacan's dogmatism, Derrida recommends that, rather than dispense with any and all distinctions between reaction and response, critics should explore the workings of the two "within the whole differentiated field of experience and of a world of life-forms."[9] Music would seem to be an apt locus for such inquiry. At first glance, responding to music in something resembling Lacan's sense would seem to demand a hermeneutic practice—that is, an interpretative or analytical attempt to divine "what it means" or "how it works." Both approaches overlap with *Problemsgeschichte*, a methodology that conceives artworks as solutions (or answers) to creative problems (or questions). In either case, music sets in motion that "second-degree reflexive power" which pushes back at immediate reactions as it weighs plausible responses. But while music can be the occasion for such reflexive and highly refined responses, it also acts on the body in multiple ways, many of them involuntary. When music is playing, feet and fingers may begin tapping without conscious instruction. Someone within earshot of music may begin to sing along, and, worst of all, end up with the tune stuck in her head. That intrusive tune, or earworm, thumbs its nose at the notion of free will, its unbidden repetitions triggered by some hidden impulse. And yet, such experiences manifest enough of an intentional dimension to prevent classification of them as reactions to stimuli on the order of jerking one's hand away from a hot burner.

While music's capacity to elicit reactions and responses involving both body and mind—with mind understood as emergent from bodily processes rather than dualistically separate from them—is currently valued not only for its invigorating or soothing results but also for its therapeutic potential, musical aesthetics historically has had a hard time accepting music's disconcertingly wide range of effects. It is not so much that critics denied music's impact on the body altogether as that they tried to limit music's value to its powers of spiritual cultivation. From Eduard Hanslick's consternation over "pathological," feeling-centered listening to Adorno's comparison of the jitterbug's dance moves to "the reflexes of mutilated animals," music's philosophical devotees have repeatedly expressed concern about the ease with which the art raises the specter of human animality by encouraging heightened physicality and indulgence in sensory pleasure.[10] Even though music achieves its effects with the help of human aptitudes scarcely in evidence elsewhere in the world, such as rhythmic entrainment via beat induction, it nonetheless has been—and in some respects, continues to be—the art that most dramatically unsettles efforts to distinguish humans from nonhumans on the basis of the mental ability to resist immediate physical impulses.[11]

This chapter revisits two influential treatises—Johann Georg Sulzer's *General Theory of the Fine Arts* (1771–74) and Hanslick's *On the Musically Beautiful* (1854)—that illuminate persistent anxieties over the admixture of reaction and response in musical listening, an admixture believed to carry with it the further threat of confusion between animal and human modes of perception and cognition. This threat became all the more urgent as humanity's "*second trauma*"—Derrida's name for Darwin's post-Copernican blow to human self-esteem—began to sink in (*On the Origin of Species* was published only five years after Hanslick's book, in 1859).[12] If humans had in fact descended from apes, and if the sovereign power of the mind over the body was the linchpin of human exceptionalism, then indulging in thoughtless sensuality via aesthetic experience might be a sure way to devolve to that distant origin. For Enlightened man, write Horkheimer and Adorno, "only perpetual presence of mind forces an existence from nature," by which they mean a distinctly human existence.[13]

From today's vantage point, however, music's formal features, social character, and inducement of empathy indicate intriguing links to the variegated, and only partially understood, domain of animal communication. The second half of this chapter considers how recent ethological research and insights into the broadly embodied nature of musicality disturb key suppositions of European musical aesthetics, lending support to views of aesthetic experience that emphasize both its human and animal dimensions. Cast in more programmatic terms, the chapter seeks to illuminate music's practically unrivaled potential to serve as a nexus where science, aesthetics, and philosophy can enter into mutually enriching conversation. Only through such multidisciplinary dialogue, I contend, can we truly appreciate the extent of music's power to ramify human existence and situate humans within the wider world of life and sound.

Sympathetic Vibrations

By suggesting that music occupies a particularly tendentious position with respect to the categories of reaction and response, I do not mean to imply that the other arts simply engage one or the other in uncomplicated fashion. Ever since its inauguration by Alexander Baumgarten, the discipline of aesthetics has been dogged by the problem of art's sensuousness, which had to be synthesized with cognition to neutralize its capacity to stimulate inappropriate reactions, especially desire. But the principles of Kantian disinterestedness or Hegelian sublation are more difficult to maintain in the face of music, which often invites (or even demands) direct physical engagement. In his commentary on music, Sulzer pointed to the ancient lineage of work songs and dance, remarking that "musical sounds themselves always imply an idea of movement," an insight backed up by recent research on music and the brain.[14] Later commentators continued to grapple with music's ability to grip the body like no other art. For Friedrich Nietzsche, the orgiastic revelries orchestrated by Dionysus, flute in hand, challenged the serene order of Apollo and his lyre.[15] Adorno framed the problem in more physiological terms, writing that "music represents at once the immediate manifestation of impulse and the locus of its taming."[16] Adorno's phrasing implies that reaction itself calls for a response—the response of taming, or, more literally, softening (*Sänftigung*). In one stroke, music illustrates the dialectical nature of the civilizing process.

What was it about music that made it so troublesome? First of all, aestheticians had long realized that music's sensory medium—sound—wielded great physical power and thus could stimulate strong reactions in listeners. "An out-of-tune note is incomparably more disagreeable and disturbing than is a clashing color," Sulzer mused. "The ear," he continued, "can be so smitten by inharmonious sounds as to drive one almost to despair."[17] Such remarks call up scenarios of fingernails on a chalkboard or metal on glass, in which the immediacy of one's physical reaction to sound crowds out the possibility of a reasoned response. Yet the immediate impact of sound, its ability to inspire involuntary reactions, also contributed to music's superior expressive force. "Nature," wrote Sulzer, "has established a direct connection between the ear and heart."[18] The mystery of this connection, marveled at by so many writers on aesthetics, could be traced in Sulzer's view to an even greater conundrum—namely, the relationship between the body and the soul, between the seat of reaction, so to speak, and the seat of response. "The aural nerves," Sulzer explained, "transmit to the entire body the impact of the shock they receive. . . . Hence it is understandable how the body, and consequently the soul, can be intensely affected by sounds."[19] Stimulation of the nerves residing in the ear appears to lead seamlessly to the humanizing responses of sentiment.

Elsewhere in the *General Theory*, however, Sulzer singled out disruptive effects much like the "shock" described in the previous paragraph as belonging to the lesser category of "accidental" versus "essential" aspects of aesthetic experience. Such effects compel rather than persuade, stun or surprise rather than move listeners, thus compromising music's mission to touch the heart by way of the body instead of the body alone. The contradiction becomes acute when Sulzer speculates that music, because of its physical efficacy and direct connection to feeling, could succeed better than the other arts in civilizing "savages." Swayed by accounts of Orpheus that describe him as animating nonhumans as well as taming wild men, Sulzer believed that music, alone among the arts, could awaken finer sentiments in the beast-like savage heart, sentiments which could then serve as the foundation of morality. But, as Matthew Riley points out, Sulzer never quite managed to distinguish that ability from aesthetic forces that short-circuit the faculty of reason—forces with which music was well endowed and that work more by inciting reactions than inspiring responses.[20] Although music appeared ideally equipped to elicit moral or sentimental responses from allegedly uncivilized listeners, it did so paradoxically through its very capacity to bring about involuntary reactions.

Despite music's evident impact on the reactive, animal body—or perhaps because of it—Sulzer took great pains to separate the art from the realm of nature in general. Dispensing with the idea that humans learned to sing by imitating birds, Sulzer linked the rhythmic regularity of music to such human activities as walking and physical labor, and he located the origin of melody in the expression of "passionate emotions."[21] But Sulzer's categories of humanity and nature begin to merge when it comes to the origin of such passions: "The individual sounds that comprise song are the expressions of animated sentiments . . . and the sentiments aroused demand to be expressed, even if against one's will, by the sounds of song, not speech. Thus the elements of song are not so much the invention of man as of nature herself."[22] At times involuntary and authored by nature, music has its origin in the physiological and affective conditions of human existence, and it evidently carries the trace of that origin no matter how refined the sentimental responses it calls forth.

Adding to an already tangled set of complications, Sulzer suggested that human music can devolve to the status of nonhuman nature. Overdeveloping the technical side of music, for instance, threatens to lead back to the animal realm from which Sulzer had detached it. Virtuosic compositions demanding great physical skill, he observed, too often come off "like a horse running in full gallop."[23] Yet Sulzer concluded that such compositions are "no more natural than Agesilaus's mimicking the song of a real nightingale."[24] The naturalness of art derives from the second nature of human creativity rather than mimetic accuracy. When the physical capacities of the performer are pushed to their limits, music becomes no more than an animal, a body devoid of soul. In a similar vein, Sulzer noted that one can write a piece of music that conforms to "mechanical rules" but lacks expression, a notion consistent with lingering views of the animal body as more like a reactive machine than a responsive being.[25] Well-crafted but inexpressive music darkly insinuated that humans were perpetually in danger of reverting to the same.

Just before Sulzer began work on the *General Theory*, the critic and philosopher Johann Gottfried Herder was also pondering the relationship between human and animal modes of expression, but with a surprising twist. In contrast to Sulzer, Herder explained the mysterious connection between ear and heart in explicitly animalistic terms. The *Essay on the Origin of Language* (1770) opens with the unforgettable sentence, "While still an animal, man already has language."[26] Herder means not specifically human language but the natural language of expression—the "screams" and "wild inarticulate tones" of early humans. These sounds, which arise automatically from the "mechanics of sentient bodies," emerge in reaction to bodily sensations and affects such as pain or fear. But it soon becomes clear that such reactions constitute the phylogenic prerequisite for response. "Even the most delicate chords of animal feeling," Herder remarked, "are aligned in their entire performance for a going out toward other creatures. The plucked chord performs its natural duty: it sounds! It calls for an echo from one that feels alike, even if none is there, even if it does not hope or expect that such another might answer."[27] This is a language "meant to sound, not to depict," a phrase that echoes Sulzer's injunction against trying to represent objects or ideas in music.[28] Animal sounds, in short, are both grounded in involuntary reactions and expressive calls awaiting a response.

Herder thought that human language originated in the power of reason or "reflectiveness" (*Besonnenheit*), which he believed to be unique to humankind. Yet he remained remarkably attuned to the kinship of all animals. In the article "On Image, Poetry, and Fable" (1787), he proposed that "metaphysics, that prideful ignoramus, ought to give up the arrogant delusion that the humblest animal is *wholly* unlike man in its activities and aptitudes, for this notion has been amply disproven by natural history. In their whole *disposition of life* animals are organizations just like man is; they merely lack human organization and the prodigious instrument of our abstract, symbolic memories: speech."[29] Although the expressivity of animal calls did not fully account, in Herder's view, for the origin of human verbal language, the close bond between affective reactions and responses lived on for him in music, a situation that, as we will see, offers another means of dismantling the opposition between Lacan's two categories.

On Musical Narcosis

By 1854, Eduard Hanslick was no longer convinced that music's moral influence was wholly positive. Whereas Sulzer had been comfortable with the notion that "music is written not for the mind or imagination, but for the heart," Hanslick, appalled by the behavior of Richard Wagner's enthusiasts, was disturbed by how Sulzer's position (and others like it) left the body at the mercy of music.[30] Believing that music amounted to an "intelligible language of sentiment," Sulzer had charted a path from the physical impact of sound to the stirring of sentiment to the cultivation of morality.[31] Hanslick, on the other hand, envisioned another trajectory for music, one that necessitated the suppression of its physical and emotional impact. Seeking to establish a scientific discipline of music aesthetics true to the aims of the post-1848 academy, Hanslick elaborated a mode of engagement with music that placed the dispassionate mind rather than the feeling body at the center of reception, thus disqualifying the body's animal reactivity from the sphere of aesthetic legitimacy.[32]

To promote his mind-centered reception, Hanslick distinguished between music's "material moment," which he considered to reside in the "natural power of tones," and its purely intellectual "artistic moment" (*OMB*, 58). Like Sulzer before him, Hanslick recognized music's peculiar power all too clearly. "The other arts persuade," he wrote, "but music invades us" (*OMB*, 50). And with invasion comes surrender, as music proceeds to overpower the nervous system, especially that of the psychologically abnormal listener.

The "unfathomable affinity" Hanslick posited between the physicality of music and the human body resembles Sulzer's "direct relationship between ear and heart," but the heart in this case has shed its broader metaphorical significance and collapsed into an all-too-material organ (*OMB*, 58). In Hanslick's view, the body reacts to music's "elemental" components—sound and impressions of motion that mimic the "dynamics of feeling"—while the mind responds to its formative aspects, including the construction of themes, phrase structure, harmony, and instrumentation (*OMB*, 58, 20). When insufficient attention is paid to these aspects, the elemental in music "shackles the defenseless feelings" of listeners, lulling them into the "passive receptivity" of "pathological" listening (*OMB*, 58). "Slouched dozing in their chairs," Hanslick jibed, these enthusiasts "brood and sway in response to the vibrations of tones, instead of contemplating tones attentively" (*OMB*, 59). Their needy bodies guzzle down arias like champagne and consider music little more than a fine digestif or an intoxicating drug (*OMB*, 60). Highlighting the body's unthinking reactivity to music, Hanslick recommended ether and chloroform as alternatives to the "effortless suppression of awareness" afforded by music (*OMB*, 59). Abusing music in such fashion puts the art in the category of natural entities and substances that take advantage of the body's receptivity but do not "make us think" (*OMB*, 60). Pleasant reactions are no substitute for the rewards of "pure contemplation," which consist in the measured responses of aesthetic evaluation (*OMB*, 58).

Pure is the operative word here, because—and this is a point on which Hanslick differs markedly from Sulzer—being moved to moral action by music constitutes a reaction just as automatic as succumbing to music's narcotic effects. If music makes us want to *do* anything, whether perform a kind act or get up and dance, then we have not acted out of "free self-determination" (*OMB*, 61). While Hanslick conceded that it would be pedantic to deny the animating effect of dance music and marches, it seems that the cultivated listener is obliged to resist such invitations (*OMB*, 54). Comparing someone moved by a musician's performance to forgive a debt to a sluggard motivated to dance by a waltz, Hanslick placed "neural stimulation" rather than a love of beauty at the root of both actions (*OMB*, 61). In a classic expression of the humanist devotion to rationality, he concluded, "To undergo unmotivated, aimless, and casual emotional disturbances through a power that is not *en rapport* with our willing and thinking is unworthy of the human spirit" (*OMB*, 61).

So much, then, for Sulzer's Orphic scenario of cultivation by way of music. Even moral uplift falls under Hanslick's axe as an essentially unwilled by-product of music's wiles. Arguing that music makes the greatest moral impact on those possessing "coarseness of mind," Hanslick remarked that "music exercises the strongest effect upon savages," an achievement that, to his mind, does the art no credit (*OMB*, 61). Ancient accounts of the power of music can be explained by the fact that humanity in its more primitive stages was "more at the mercy of the elemental" (*OMB*, 62). In this respect, Hanslick implied, humans used to be more like animals. After repeating decades-old (and rather questionable) reports of the music-inspired feats of animals, Hanslick asked, "is it really so commendable to be a music lover in such company?" In contrast to animals and "savages," modern Westerners "cherish a contemplative kind of pleasure in the products of music art which paralyses [*paralysirt*] music's elemental influence" (*OMB*, 63).[33]

Paralyzes music's elemental influence. Put away the ether, listeners, and grab the Novocain (synthesized by a German chemist in 1905, one year after Hanslick's death). If you wish to be civilized, ignore your body when listening to music, shut down the affective mechanisms responsible for transforming musical sound into subjective feeling. Such an achievement, if that is what it is, depends on channeling physical sensations into the safer territory of the appraising intellect. Midway through his treatise, Hanslick promises not to devalue the sensuous, as had "older systems" (such as Hegel's) that stressed art's moral import or the ideas it conveyed (*OMB*, 29). He even claims to see no problem in taking "naïve" pleasure in music's "merely sensuous aspect," so long as one does not allow the unseemly translation of sensations into feelings (*OMB*, 60). But what Hanslick really wanted to establish was a direct path leading from sensations to the auditory imagination, where the mind contemplates the "rich variety of the succession of sounds in itself" (*OMB*, 60). Music's tonally moving forms, as the art's only content and source of its beauty, may indeed resemble the dynamics of feeling, but they must not be taken as spurs to any specific feelings (*OMB*, 29). From the standpoint of reception, any straying from the conduit joining sensation and intellect to the realm of feeling ends up, for Hanslick, in the dead end of pathological listening. In sum, Hanslick's treatise constructs an imaginary body for whom objective sensations lead to equally objective cognitions without stirring up any subjective feelings along the way.

The aesthetic doctrine endorsed by this imaginary body became known as formalism, which, in its more conservative guises, demands that confrontations with (and discourse about) artworks concentrate exclusively on constructive elements. While there is nothing wrong with willingly focusing on such elements, it is hard to shake the impression that the body of Hanslick's formalist listener arises not out of convincing aesthetic or physiological realities but out of objectivist disavowal of the reactive, affective body's contribution to musical experience. Hanslick knew full well that nearly everyone experiences some sort of feeling when listening to music. The searching questions his treatise asks about how "the sensation of tone becomes feeling or mood" show that this was indeed one of his primary concerns (*OMB*, 54). But he did not believe that physiology was prepared to answer such questions, which again boil down to the intractable problem of "how the body is connected to the soul" (*OMB*, 56). Hanslick's formalism, then, arises out of an injunction. What one cannot know the causes of, his argument goes, one should not talk about. Better simply to describe (if description is ever simple). But this does not change the fact that formalist descriptions of music are discursive constructs sitting atop an abyss—the abyss of bodily reaction and response.

Musicking Multiplicities

After giving modern musical formalism its decisive impetus, Hanslick abandoned his objectivist project in disillusionment, returning instead to the more journalistic idiom of music criticism.[34] Hanslick's about-face, however, does not mean that all tenets of formalism should be discarded, particularly not its objections to the claim that music's primary task is to represent emotions. Music, for Hanslick, did not represent anything; instead, it transposed the dynamics of feeling and movement into an acoustic, artistic medium. As I argued in chapter 2, focusing on music's dynamic qualities, its semblance of growth and purposeful motion, yields a view of musical works not as idealized formal structures but as akin to living things. Despite his intuitions along these lines, Hanslick drew the unfortunate conclusion that music at its most cultivated should serve exclusively as an object of contemplation rather than a stimulus to feeling or physical action, a position just as limiting as the theory of representation he tried to refute. Complementing a formalist understanding of what music does with a physiological and existential conception of music's manifold "doings" helps clear a path beyond the intellectual biases of Hanslick's treatise and its scholarly descendants.

This path, incidentally, stretches both forward and backward. Goethe's brief sketch for a "Theory of Tone" (1810), for instance, presents a more straightforward interpretation of music's organic nature than those of his Romantic contemporaries, in that Goethe was concerned not with the organization of musical works but with the vocal and bodily prerequisites of music making. The sketch divides the phenomenon of tone into organic, mechanical, and mathematical aspects. Whereas the latter pertain to the construction of instruments and acoustics, respectively, the organic category encompasses the physiological and subjective factors involved in musical experience. The first member of this category is the voice, a phenomenon that extends beyond the human to include "the voices of animals, especially birds." Next comes the return of sound through the ear, an occurrence by which "the entire body is stimulated."[35] Finally, Goethe proposes the study of "rhythmics," or the broader physical involvement in music. Activities such as marching and dancing, he observes, are forms of human behavior that overlap with "all organic movements," especially those that "manifest themselves through diastoles and systoles."[36] Goethe's remarks on the organic qualities of music suggest a rewriting of Adorno's quip: "With human means art wants to realize the language of what is not *only* human."[37] Despite its brevity, Goethe's sketch telegraphs the immense scope of his unrealized project, the organic part of which has been taken up in recent years by researchers in biomusicology (the study of the physiological preconditions and effects of musicking on the human organism) and zoomusicology (the study of patterned sound production in animals).[38]

From the standpoint of human physiology, music's "doings," in the sense alluded to earlier, turn out to be remarkably complex, a fact that undermines formalist efforts to restrict listening experiences to the intellectual appreciation of constructive elements. Far from being strictly contemplative or even solely auditory, music engages multiple senses, parts of the body, and regions of the brain.[39] Building on theories of embodied cognition, music psychologists Jessica Phillips-Silver and Laurel J. Trainor argue that "sensory perception cannot be separated from the multisensory experience of our bodies," an imbrication that has special consequences for the understanding of musical rhythm.[40] Phillips-Silver's many studies of rhythm and bodily movement have shown that auditory, motor, and perhaps even vestibular systems work together in the perception and production of music.[41] Her work, along with that of other like-minded researchers, serves as a much-needed corrective to the traditional focus on narrowly auditory phenomena in music cognition studies, a focus that has obscured what Petri Toiviainen and Peter E. Keller call the "corporeality and multimodality of music processing."[42] The problem is that, as Peter Sedlmeier, Oliver Weigelt, and Eva Walther observe, "contemporary psychology of music has been concerned almost exclusively with the *mind* behind musical experiences," even though people in all cultures "move their bodies to the rhythms of music in a variety of ways."[43] It is as if Hanslick were continuing to direct research programs behind the scenes. Encouraged by the ubiquity of music's allegiance to movement and dance, these three researchers advance the hypothesis that "music co-evolved with body movements and that this relationship is fundamental to our nature."[44] Herder was clearly on the right track when he hazarded that "we hear almost with our whole body."[45]

Empirical studies of music's corporeal impact are ripe for harvesting in the more sociological and interpretive settings of musicology and ethnomusicology. In her work on musical trancing, Judith Becker situates musical experience within "multiple senses of embodiment." These senses include the body as a "physical structure in which emotion and cognition happen," the body as "the site of first-person, unique, inner life," and the body "as involved with other bodies in the phenomenal world."[46] In an effort to honor this multiplicity, Becker weaves together an account of musical reaction and response that is "biologically based, psychologically sophisticated, and attuned to cultural nuances."[47] The varieties of music-induced arousal of the autonomic nervous system alone are sufficiently striking: "shivers, goosebumps, changes in breathing and heart rate, tears, weeping, [and] changes in skin temperature." These reactions, Becker contends, "precede language and evaluation," and their involuntary nature is not separable from the aesthetic responses they accompany.[48]

Although Becker maintains that the emotions stirred up by music are "rooted in basic physiological arousal felt in the body," the relationship between the two is unlikely to be linear.[49] Based on a pair of studies, Thomas Schäfer and Peter Sedlmeier surmise that the self-reported emotional arousal of listeners (which corresponds to Becker's "first-person" sense of embodiment) does not always correlate with physiological arousal (that is, Becker's body as a "physical structure" subject to measurement).[50] They propose that music preference is (perhaps not surprisingly) more closely related to the emotional rather than physiological arousal it brings about. Even here, though, complications abound. "We should not assume," writes Tia DeNora, "any simple one-to-one correspondence between musical genre, style or acoustical features and psycho-social or even physiological musical response."[51] Arnie Cox nuances this claim, arguing that music preference depends in part on whether listeners like or dislike what music invites them to do—that is, on the overt or covert mimetic responses music elicits.[52] These responses include what Cox calls exertion schema—roughly, imagined movements to music, whether performance-related or cross-modal—and imagined or real vocalizations. Cox's conclusions indicate that the highly refined and culturally conditioned cognitive processes central to music appreciation (with "cognition" here meaning integrated mind–body processing) operate largely unconsciously, supporting Derrida's claims on behalf of the "reactionality in the response."

The intricate and complex modalities of musical pleasure and embodiment to which these and other studies testify lend empirical support to Nietzsche's speculative recasting of the "subject as multiplicity."[53] This is a subject for whom art excites manifold regions of the human organism susceptible to pleasure. *The Will to Power* describes art as stimulating the "feeling of life" and as a route to intoxication that respects no distinctions between mind and body.[54] Music, for Nietzsche, has very little to do with calling forth the kind of linguistic responses that interested Lacan and very much to do with channeling bodily reactions toward a state of power-drenched "perfection."[55] In opposition to Hanslick, Nietzsche viewed the body's reactivity as an active rather than passive force, one that generates feelings of intensity, plenitude, and overflowing sensuality in the face of aesthetic encounters. Scorning the "absurd overestimation of consciousness" that underwrites human exceptionalism, Nietzsche cast aside the prejudice according to which any embrace of the unconscious or involuntary aspects of existence is denigrated as "becoming animal."[56] The philosopher instead described the reception of art as a thoroughly animalistic affair, and art itself as a reminder of "states of animal vigor."[57]

Gilles Deleuze and Félix Guattari pick up where Nietzsche left off, advocating for a "becoming-animal" that consists not in imitating animals but in "composing a body with the animal."[58] This body is by definition multiple, eluding any preconceived idea of what a body should be. Becoming-animal and becoming-plant are molecular rather than molar affairs; they thrive on particularity and multiplicity rather than "the imitation of a subject or a proportionality of form."[59] Of course, human bodies are already animal bodies; "there is little foundation," Deleuze and Guattari aver, "for a clear-cut distinction between animals and human beings."[60] The task is to recognize one's own states of animal vigor (in Nietzsche's words) as the gateway to dissolving the self-identical subject. Music, which "takes as its content a becoming-animal," has a special role to play in this regard. Orchestral music in particular constitutes an unparalleled vehicle for the exploration of multiplicity and for the transformation of the singular subject into a molecular collectivity. In Elizabeth Grosz's formulation, such music may be heard as a sonic analogue of the "microsubjectivities" inhabiting every living thing. "Each complex organic body," she writes, "is composed of a multiplicity of subjects or consciousnesses, a multiplicity of organs that, like musical instruments, function in harmony with the body's mnemic themes."[61] If the bowing, blowing, and beating of instruments enact the becoming-animal of human voices and gestures, the reverse is also true, in that music provides a venue for the becoming-human of animal sounds (Beethoven's cuckoo and nightingale, Olivier Messiaen's orchestral aviary). Deleuze and Guattari suggest that "the horse, for example, takes as its expression soft kettledrum beats . . . and the birds find expression in *gruppeti* [sic], appoggiaturas, staccato notes that transform them into so many souls."[62] The rhythms of galloping aside, it would appear that song and stridulation are top contenders for supremacy in the department of musical becomings. "All music is pervaded by bird songs," Deleuze and Guattari wager, but contemporary music, with its "chirring, rustling, buzzing, clicking, scratching, and scraping," trades avian melodiousness for the pleasures of becoming-insect.[63] With its polyphony of not-quite-human, not-quite-animal sounds, orchestral music confounds singular identities, luring us to become another, to become something (or many things) other than what we (think we) are.

Aesthetics across Species Lines

"Art begins with the animal," Grosz contends, upsetting Herder's more delicate positioning of beauty "between the angel and the animal, between the perfection of the infinite and the sensuous, vegetal gratification of cattle."[64] Crossing Darwin with Deleuze in a virtuosic feat of philosophical husbandry, Grosz portrays the feelings of intensity inspired by art as explicitly erotic. "Art," she argues, "is the sexualization of survival or, equally, sexuality is the rendering artistic, the exploration of the excessiveness, of nature."[65] Experiences of art, as well as artworks themselves, are, in Grosz's reading, varieties of overabundance that both derive from and feed the body's erotic energies, and in this regard, they resemble performances in the animal world that go beyond sheer necessity by carrying out complex protocols of sexualized display.

Music's origins, however, likely spoke to an array of social needs, not just those of sexual congress.[66] An embodied conception of musical experience in which reaction and response flow into and support one another opens out onto the broader field of animal communication in a manner productive for aesthetics after humanism. Consider once again Herder's account of animal sounds as both the product of bodily reactions and bids for a sympathetic response. Revisiting this theme thirty years after his origin of language essay, Herder argued that a human being is able to "lend his sympathy to every aroused being whose voice reaches him," while animals respond only to the sounds of their own species.[67] Music, whose tones arise from the physical excitation of resonant bodies, is one such voice: "the *answering voice* of the one who is feeling," a phrase that recalls his reference to animal sounds as seeking "an echo from one that feels alike."[68] With these words, Herder comes close to construing human affect as a silent call to which music is primed to respond. When listeners "answer" music with yet another round of feeling, they react to music's sonic power and respond to it as the voice of an other, as the acoustic representative of other living beings. The most authentic response to music, Herder suggests, is the desire to join in. "Sympathy" is Herder's name for the fusion of reaction and response that permeates the aesthetics of living sound, including music.

Herder's notion of music as a special kind of voice indicates that the technique of call-and-response is more than just a stylistic trait. Instead, call-and-response is implicit in all acts of music making, and in this respect, music shares an affinity with a wide range of animal behaviors. Many animals engage in vocalizations meant to be recognized and responded to by others, although these others do not necessarily occupy the position of Lacan's Other. Numerous species deploy "contact calls" whose unique character serves to keep a group or family together. Some animals, such as elephants and meerkats, use distinct calls to indicate current location versus the intention to move, while disc-winged bats native to Costa Rica emit separate "inquiry" and "response" calls during roosting.[69] Flying bats of this species can distinguish between calls belonging to members of their own or another group, an ability that helps them choose where to roost for the day. Such behavior far exceeds the narrow scope of involuntary reactions and may even resemble practices that were essential to the origins of human music. Thomas Geissmann's work on gibbon songs proposes that apes and humans shared an "ancestral form of loud call" that served to define territory, intimidate others, broadcast location or announce danger, and promote group cohesion. "Probably the most likely function of early hominid music," Geissmann asserts, was to "reinforce the unity of a social group toward other groups," a purpose still discernible in hymns, military marches, sporting songs, and allegiances to individual musical styles.[70]

That's all well and good, a Hanslick-style formalist might retort, but functional music, with its baggage of vestigial animality, does not represent the most cultivated sort of music, which is predicated on the notion of autonomy from social imperatives. Yet the points of contact between human and animal song are not exhausted by their social functionality. Ethologist Peter Marler has proposed the term "phonocoding" to describe the recombinant vocalizations of birds and whales. Unlike calls that transmit information such as location or impending danger, phonocoding involves stringing together sound patterns in differing orders by learning from the songs of others. As Marler explains, "animal songs that are learned and that depend on phonocoding for signal diversity are, like human music, primarily nonsymbolic and affective."[71] Given Hanslick's recognition that music regularly inspires feelings in listeners (although he did not believe such feelings should be a matter for learned discourse), music under his formalist description could very well be deemed nonsymbolic and affective.[72] Marler's conclusions confirm that formal manipulation of sound patterns is not an exclusively human ability. Indeed, his description of the songs of male winter wrens stirs thoughts of the recurring melodic and rhythmic motives of human music: "Each song in the repertoire contains phrases drawn from a large pool that recur again and again, but in each song type they are arranged in a different sequence. Evidently what happens when a young male learns to sing is that he acquires a set of songs from the adults he hears and breaks them down into phrases or segments. He then creates variety and enlarges his repertoire by rearranging these phrases or segments in different patterns."[73] While such songs serve to identify the singer and may benefit the animals in other ways, they do not appear to serve functions typical of "lexicoded" vocalizations, which encode meaning into individual vocal units. The abundance of birdsong repertoires, Marler proposes, generates "sensory diversity" rather than "enrich[ing] meaning" in a linguistic sense.[74]

The processes of learning that underlie both human and avian vocal production also display certain similarities, likely as a result of the particular genes and brain structures these species share (despite the disparate histories of their recruitment).[75] Like humans, birds are capable of both original invention and imitation of others. Björn Merker describes several paths songbirds take toward mature song, one of which is simply learning to reduplicate the songs of adult members of its species.[76] This process moves through several stages, until the young bird correctly reproduces its species' characteristic call. Merker equates the kind of imitative learning on display in such cases with the transmission of a cultural ritual.[77] Other birds' repertoires remain open-ended into adulthood, which means that adults continue to assimilate or invent novel song elements over the course of their lives. Mockingbirds mimic the sounds of birds such as nuthatches and whip-poor-wills with near-perfect accuracy, while the marsh warbler may incorporate the sounds of a hundred species into its calls.[78] Original invention, finally, occurs by way of variation and recombination of learned elements in the manner described by Marler. Either "the sequencing of a set of song elements is varied on a nondeterministic basis," Merker explains, or "model patterns are disassembled into phrases and fragments and reassembled into new unique song types."[79]

The growing body of work on animal communication and physiology suggests that Derrida's skepticism regarding Lacan's separation of human response from animal reaction should also be directed to certain distinctions between human music and animal sound, especially those that turn on specious conjectures regarding autonomy versus functionality. In one of the bolder passages of his essay, Marler encourages readers to hear the repertoires of phonocoding birds "as providing aesthetic enjoyment or as alleviating boredom in singer and listener."[80] Given subsequent research which has shown that singing releases endorphins and opioids in the brains of birds, perhaps Hegel was not entirely wrong to suggest that birds sing for the "immediate enjoyment of self."[81] Philosopher and ornithologist Charles Hartshorne insisted on what he called the "aesthetic" factor in birdsong, and he marveled at the "astonishing" degree of intelligibility the vocalizations of birds have for humans, at least in some cases.[82] Employing a robustly analogical style of thinking, Hartshorne cited birdsong equivalents of accelerando and ritardando, crescendo and decrescendo, interval inversion, simple harmonic relations, and variation and recombination of thematic material. Alongside these recognizable features are plenty of song elements that human ears find difficult or impossible to parse, such as the extremely rapid notes of the winter wren (up to thirty-six per second) or the multiphonics of mockingbird and canary songs.[83] The acoustic and formal territories of birdsong and human music overlap but are not coextensive, which is hardly surprising given their divergent evolutionary trajectories.

One can marvel at the similar outcomes of independent evolutionary processes (referred to as *convergent evolution*) without simply assuming that human and animal song arose from the same functional, expressive, or aesthetic imperatives. Even as Grosz acknowledges that "there is no direct line of development between birdsong and human music," she contends that both kinds of song represent "the opening up of the world itself to the force of taste, appeal, the bodily, pleasure, desire."[84] By contrast, Gary Tomlinson cautions against the "feel-good anthropomorphism" that would place, for example, an Italian aria and a robin's melodious call into the all-embracing category of "song."[85] Tomlinson's placement of scare quotes around this word when referring to the vocalizations of nonhumans evidently arises from a desire not to impose a specifically human conception of song onto other beings. Yet by only grudgingly extending the cognitive capacities involved in music making to other species, Tomlinson reveals the recognition of difference to be, in this case, the flip side of human exceptionalism.[86] The modular structure of some bird and whale songs suggests that these animals are capable of the combinatorial cognitive processing Tomlinson places at the center of his evolutionary history of human musicality. These resemblances raise the tantalizing prospect of a cross-species formalism based on comparative study of the pattern-forming vocalizations and percussive behaviors of animals, humans included.[87]

As in the case of human music, however, a formalist approach cannot by itself account for the full biological, affective, and phenomenological significance of animal communication. Philosopher Kathleen Marie Higgins underscores the role of empathy—a concept not unlike Herder's notion of sympathy—in the human reception of nonhuman sounds. Proposing that the vocalizations of some animals can be considered music, she writes, "Just as we recognize the life and energy of other human beings when we listen to music, we recognize kindred life and energy of birds and other creatures through the sounds they produce. The delight we take in birdsong, for example, is continuous with our pleasure in human music, for it is similarly grounded in a recognition that we are part of the same living world."[88] But empathy too only takes us partway into the living worlds of other animals. Although biologists confidently identify particular calls as being related to territory, mating, or alarm, Higgins recommends a healthy agnosticism: "We don't know whether or not the meaning [of song] experienced by the animal is monodimensional. Affective meanings such as joy, self-expression, responsiveness to some change in the environment, and interactive engagement with other animals and their sounds are all levels of meaning that might be experienced by a vocalizing animal."[89] Where knowledge of these meanings is lacking, formalist analysis, comparative biology, and speculation driven by empathy are all resources that can help us map connections and disjunctions between human and nonhuman modes of embodied communication.

The power of empathetic listening to stimulate human creativity has itself been the object of considerable speculation. Despite Sulzer's skepticism, many thinkers (including, as we have seen, Deleuze and Guattari) have entertained the possibility that the history of human music is partly the history of cross-species encounters. The Roman author Lucretius famously suggested that music originated in the imitation of birdsong, and this idea, while unprovable, persisted for many centuries.[90] The anonymous *Kurtzgefaßtes musicalisches Lexicon* of 1749, for example, includes a passage by Wolfgang Caspar Printz, who counted birdsong among the stimuli of musical invention, "for it is plausible that, in their idle hours, people wanted to reproduce it through mimicry."[91] Several decades later, the versatile author John Hawkins, taking cues from both Lucretius and Athanasius Kircher, held that "the voices of animals, the whistling of the winds, the fall of waters, the concussion of bodies of various kinds, not to mention the melody of birds, as they contain in them the rudiments of harmony, may easily be supposed to have furnished the minds of intelligent creatures with such ideas of sound as time . . . could not fail to improve into a system."[92] And in 1819, the English music historian Thomas Busby maintained that "the notes of birds, as a living melody, a melody not subject to chance, but no less constantly than agreeably saluting the sense, could not but excite human imagination."[93] Whatever one thinks of such claims, their authors do at least imagine a world in which early humans were responsive to, and creatively inspired by, the sounds around them. Indeed, the Paleolithic hunters and gatherers who feature in Tomlinson's evolutionary story must have been highly attuned to their auditory scenes, which were full of acoustic signs of danger, beneficial resources, and the presence of others. Although Tomlinson acknowledges this, his theory of music's cognitive prehistory leaves little room for either comparing the physiological mechanisms of vocal production across species or considering the possible influence of human interaction with the larger world of sound on the development of music.[94]

If the notion of human music's indebtedness to birdsong is ultimately a myth, it is, as Matthew Head has suggested, an especially fertile one.[95] It is also timely, in that the growing recognition that a viable human future on this planet must also be a multispecies future makes the prospect of thinking about humanity's historical entanglements with nonhumans all the more instructive. The myth of music's origins in birdsong gains added resonance in the face of certain relics dating from late Paleolithic times. Denizens of Europe in the last ice age carved images of animal and plant-like forms onto bones, antlers, and ivory. Sometimes these artifacts display a redoubling of image and material template: pictures of deer etched onto deer antlers, a whale engraved on a piece of whale ivory.[96] Might the bone flutes that provide the earliest material evidence of music making engage in a similar redoubling, such that blowing into a flute made of swan or griffon vulture bone produced sounds understood to resemble those of birds? Could such inventions speak to the gradual emergence of discrete pitch production and perception not only out of human protolanguage, as Tomlinson argues, but also out of human audition and imitation of pitches resounding in the wild?[97] The first songbirds originated more than fifty million years ago, so birds had plenty of time to refine their craft before hominins appeared on the scene. If our reactive responsiveness, or responsive reactivity, to the sounds of nonhuman others turns in part on our capacity for empathy, then empathetic listening and its affective freight are likely to have both deep roots and broad outcomes, of which music may be only one. Another outcome, surely, is the acoustic biophilia (to invoke E. O. Wilson's term) that occasions human pleasure in the songs of birds. Here, however, as in so many other deep historical matters, we pass beyond the reach of the sciences and the humanities alike, leaving us to wonder whether human musicking and human biophilia—or, more broadly, creativity and affectivity—have a shared story to tell.

6

On Not Letting Sounds Be Themselves

"It began with birds," states John Luther Adams in "Resonance of Place" (1994), an essay chronicling the development of his compositional philosophy and technique. Spanning twenty years of musical output, the essay recounts Adams's early efforts to translate, echo, or evoke bird songs—"those marvelous languages which we do not speak and which we may never fully understand"—and other features of the Alaskan wilderness in his music.[1] Adams shows how in the interval between the instrumental miniatures *songbirdsongs* (1974–79) and the multimovement epic *Earth and the Great Weather* (1993), he gradually supplanted the imitation of nature with what he calls "sonic geography," with music that "*is* landscape," music that "conveys its own inherent sense of place."[2] As a corollary to this project, Adams argues that "attentive listening to wild sounds" can "expand our understanding of musical meaning."[3] Adams's statement inverts the more frequently encountered claim that the aesthetic sensibility listeners bring to music can enrich their appreciation of environmental sounds, whether natural or human-made. In the same essay, for example, Adams praises John Cage and other compositional forebears for encouraging listeners to hear "the entire world of sound as music," a view that has long attracted enthusiastic proponents.[4]

In this respect, "Resonance of Place" is a typical product of that strand of modernism which seeks to erode the boundaries between art and nature, expand the sonic resources available to composers, and heighten attention to the sensuous qualities of sounds.[5] Since John Cage's *4′33″*, listening to environmental sounds and listening to music have ceased, at least for those whose taste runs to panaestheticism, to be fundamentally different endeavors. Yet Adams's essay also insists on distinguishing rather than conflating the sounds of nature and the sounds of music. While music has traditionally been understood in terms of its expressive meanings, Adams contends, "Sounds as they occur in the world simply *sound*. Their greatest power and mystery lie in their direct, immediate and non-referential nature. If we listen carefully enough, occasionally we may simply hear them just as they are."[6] Adams's language evokes that of Cage in the essay "Experimental Music" (1957), which urged composers to go about "discovering means to let sounds be themselves rather than vehicles for man-made theories or expressions of human sentiments."[7] By basing sonic geography on the prototype of natural sounds being "just as they are," Adams effectively appropriates and repurposes Cage's famous injunction to transfer "nature's manner of operation into art."[8] The upshot seems to be that aural encounters with the natural world help listeners refine their capacity for "direct" or "immediate" apprehension of sound. Composers, in turn, can find inspiration in nature to create music that is concerned not so much with expression as with pure acoustic qualities.

But something about this argument does not satisfy. Aside from the confusion in "Resonance of Place" about whether natural sounds are to be heard "as music" or "just as they are"—or are these options somehow equivalent?—one might question whether the modernist concept of immediate sound, or, more poetically, of letting sounds be themselves, really brings contemporary music closer to the "wild sounds" of nature. What semiotic principles, musical or otherwise, would justify such a claim? Benjamin Piekut observes that for the Cage of "Experimental Music," "nature is figured at its most traditionally modernist—that is, as raw sound."[9] This association, Piekut explains, derives from a modern Western understanding of the world, memorably diagnosed by Bruno Latour, that separates nature and humanity into two opposing camps and reserves practices of meaning making for humans.[10] Determined to burst the traditional confines of those practices, Cage accorded nature the role of austerely indifferent legislator (or, less generously, scapegoat) of aesthetic experimentation.

As Piekut's analysis implies, the idea that nature is the domain of direct, immediate, or raw sound is of relatively recent vintage, a fact that raises questions about the aesthetic outcomes of attempts to compose with sounds that "simply sound." Do musical works composed in accordance with a view of nature as semiotic blank slate succeeding in being more "natural" than the varieties of nature-themed music predating modernism—music like, say, Robert Schumann's 1849 piano cycle *Waldszenen* (*Scenes of the Forest*), op. 82? Schumann's cycle is generally considered to belong on the popular, or *Biedermeier*, end of the Romantic musical spectrum, which is to say that it is in no way protomodernist. Its nine accessible movements imagine the forest as a space of both human and nonhuman habitation rather than as a pure sylvan wilderness. Does this mean that *Waldszenen*'s musical poetics stand at a further remove from "nature's manner of operation" than those of Adams or Cage?

This chapter approaches the question not by staging a contest between Schumann's music and that of later composers but by considering how the semiotics of "sounds themselves" departs from both familiar conceptions of music and contemporary perspectives on nonhuman semiosis. To that end, I place Adams's and Cage's polemics, as well as several commentaries on composing with natural soundscapes, in dialogue with recent ventures in Peircean semiotics by Gary Tomlinson and Eduardo Kohn.[11] Tomlinson has done more than any other musicologist to bring biosemiotics into the discipline's orbit, and, although the pages that follow take issue with several of his arguments, his work nonetheless illuminates enigmas of signification, musical and otherwise, that few others have confronted. This chapter elaborates on those enigmas as they play out in contemporary "sonic geographies" by Adams and the Norwegian composer Jana Winderen as well as the movement "Vogel als Prophet" ("Bird as Prophet") from Schumann's *Waldszenen*. I argue that viewing the natural world as the rightful home of sounds themselves—of essentially meaningless sounds fundamentally different from the meaning-laden sounds of humans—works against the desire, so often expressed by advocates of such views, to close the distance between human music making and nonhuman sound making. Reintroducing meaning into the full spectrum of semiosis, by contrast, provides an alternative means of both closing and maintaining that distance, even as it sets limits on the human ability to make such determinations.

The Sound of Signifying

Let us return to the question of how Adams thinks attentive listening to wild sounds can expand our understanding of musical meaning. In brief, the composer argues that when humans listen to sounds in nature, they employ a mode of hearing that predates the development of communication via symbols, whether linguistic or musical. When this kind of hearing predominates in musical experience, Adams suggests, a more primordial way of being can flourish within the civilized domain of culture. Listening with scant concern for meaning or message, we become aware of "those profound and mysterious connections between the sounds we make and the larger, older world."[12] Adams's remarks resonate with the perennial intuition that music occupies different semiotic terrain than language. Cage, for example, explicitly contrasted the experience of listening to new music, and presumably to natural sounds as well, with interpreting the sonic symbols of language. The new hearing, he proposed, was "not an attempt to understand something that is being said, for, if something were being said, the sounds would be given the shapes of words."[13] The direction of Adams's argument in "Resonance of Place" is thus all the more surprising: "Human music is generally a symbolic and a semantic phenomenon, in which the relationships *between* sounds mean more than the sounds themselves. But sounds as they occur in the world are not symbols, subjects or objects. Inherently, they do not represent or evoke anything other than themselves. They simply *sound*. Their greatest power and mystery lie in their direct, immediate and non-referential nature. If we listen carefully enough, occasionally we may simply hear them just as they are." Following these precepts, Adams has sought to devise compositional strategies that combine the "symbolic strictures of musical semantics" with the irregular temporal flow of natural soundscapes. The sonic geographies that result invite what Adams calls a "non-metaphoric" style of listening.[14]

If Adams's ascription of symbolic reference to music may bring readers with more "drastic" inclinations up short, his corresponding denial of referentiality to the nonhuman sounds of nature is even more perplexing.[15] Is it really true that sounds such as "the primal music of bird songs and animal cries, the voices of wind and water"—is it really true that these do not, as he says, "represent or evoke anything other than themselves"?[16] Adams's notion of natural sounds as inherently nonreferential, as sounds that "simply sound," remains oddly insensitive to the signifying potential of acoustic phenomena. Part of the confusion, perhaps, arises from the fact that reference is usually understood on the model of language, where words normally refer to something quite different from the sequences of sounds that comprise them. According to Kohn's interpretation of Peircean semiotics, however, reference is not (always) a matter of signs standing for something else entirely but of signs standing "for something in relation to a 'somebody.'"[17] The smell of a mouse, for instance, stands to a snake for a source of food, but what the smell stands for is not separable from the mouse in the way that the sounds making up the English word *mouse* are separable from the (generalized) creature to which they refer.

If the signs interpreted by animals are referential in this manner, then semiosis serves the purposes of flourishing and survival—giving the lie to Cage's conviction that, in the words of Lydia Goehr, the sounds of nature supply the paradigm for "existing without purpose."[18] Animal sounds, as Herder recognized, are made for other animals, some of whom are the intended recipients while others are not. "What is a cry," ask Gilles Deleuze and Félix Guattari, "independent of the population it appeals to or takes as its witness?"[19] From a Peircean standpoint, animal calls signify primarily by way of iconicity and indexicality: they are recognizable as being like the sounds made by an individual or species and as utterances that indicate the presence, condition, and intentions of a sound-making creature. As Kohn writes, iconicity and indexicality are "representational modalities shared by all forms of life."[20] Even though wind and water may not make sounds for the purposes of communication, those sounds can still function as signs for sentient observers—as indices, say, of an oncoming storm or a source of refreshment. Any sound that attracts the attention of an observer can be either an index, in that it points to some dynamic thing responsible for making the sound, or an icon, whose similarity to (or lack of difference from) some other sound is noticed. Such sounds do not "simply sound" but convey meaningful information about the world. Adams concedes as much in the essay "The Place Where You Go to Listen" (1997). The essay tells the story of a skilled listener in the Alaskan wilderness, a woman who "heard small voices whispering: 'I am *uqpik*. I am river willow. I am here.' 'I am *asiaq*. I am blueberry. I am here.'"[21] Even those not capable of hearing the speech of a blueberry bush can still hear the wind striking its branches or an animal rustling within it, indices that help observers construct "auditory scenes," or sonic representations of place.[22]

Glossing Peirce's third type of sign, Kohn explains that symbols "refer to their object indirectly by virtue of the ways in which they relate systematically to other such symbols."[23] While symbolic systems have been developed to their furthest extent by humans—as far as we can tell, anyway—a substantial quantity of ethological literature has documented the at least minimally symbolic function of certain animal calls. Peter Marler observes that nonhuman vocalizations were once considered purely affective, in keeping with the assumption that only humans are capable of symbolism.[24] Ethologists have since shown that animal calls can convey information about location, movement, and impending danger in a fashion that borders on the symbolic, in that meanings arise out of systematic differentiations between sounds. In an activity Frans de Waal calls "referential signaling," the vervet monkeys of Kenya employ a number of distinct alarm calls, each corresponding to the type of predator spotted in the vicinity.[25] Even if such phenomena do not constitute full-blown symbolic systems, they do at least illustrate that animals who heard only "sounds themselves" in the utterances of their companions or rivals would not last very long.

What, then, is a sound itself, a sound "just as it is," and why have composers and listeners since at least the mid-twentieth century been so eager to ascribe that status to natural sounds? The concept has semiotic, philosophical, and technological implications, corresponding roughly to how Adams, Cage, and theorists of musique concrète have formulated it. For Adams, sounds that just sound have no semantic or symbolic meaning; for Cage, they have no purpose. For Pierre Schaeffer and later exegetes, such sounds are divorced from their sources with the assistance of recording technology and appreciated for their acoustic properties.[26] In all these cases, a preoccupation with sounds themselves says a lot more about the nature of particular listeners than it does about the sounds of nature. In short, not recognizing the semiotic character of natural sounds is a mark of one's distance from nature. At one time, all humans depended for their survival on apprehending the world as a panoply of signs. The beneficiaries (and victims) of modernization and industrialization, by contrast, rarely need to locate running water or prey to survive, and they (we) depend on technologically equipped specialists to grow and deliver food, predict storms, and identify other salient environmental patterns. Indeed, the contemplative attitude toward ambient sounds that works like *4'33"* sought to cultivate was, as Richard Taruskin has pointed out, not handed down from nature but the product of Western aestheticism.[27] This legacy is still evident in the writings of latter-day Cage enthusiasts such as David Rothenberg, who asserts that "music in nature is any series of sounds that can be appreciated for their depth, beauty, and artistry." To be musical, he continues, a natural sound need only be heard "as a beautiful form that can be enjoyed in itself apart from its purpose in the world."[28] Although this view would seem to be modulated in Rothenberg's later study of the arbitrary rather than purposeless character of aesthetic traits selected for in the wild, it is worth recalling Dario Martinelli's observation that although the biological register of the aesthetic is not "totally utilitarian," it is also not "totally useless."[29]

Taruskin offers a compelling alternative to the customary view that Cage derived the practice of letting sounds be themselves from Eastern meditation, yet one might also recognize the influence of modern information theory, in which what matters are the physical differences a signal encodes (and the challenges to its accurate transmission and decoding) rather than its meaning. In a foundational text, Claude Shannon writes, "The fundamental problem of communication is that of reproducing at one point either exactly or approximately a message selected at another point. Frequently the messages have *meaning*. . . . These semantic aspects of communication are irrelevant to the engineering problem."[30] Translating this attitude to the musical sphere means that the listener, modeled on a tape recorder or oscilloscope, relinquishes concern with what music might mean and instead focuses on a shifting continuum of sonic differences.[31] Cage tries to brighten what seems like a bleak scenario of immersion in auditory stimuli stripped of significance by positing a connection between deracinated sounds and human affect and imagination. "Hearing sounds which are just sounds," he muses, "immediately sets the theorizing mind to theorizing, and the emotions of human beings are continually aroused by encounters with nature."[32] But if this is so, it is because sounds in nature are not just sounds but signs—signs, or indices, of presences in the world, signs that establish physical, affective, and interpretive relationships between listeners (whether human or not) and their surroundings. Cage alludes to this broader semiosis when he states, in veritable Romantic fashion, that "trees, stones, water, everything is expressive."[33] Expressive of what? Of presence, of the "I am here" of everything.[34]

Cage's and Adams's genuine fascination with the natural world makes odd bedfellows with the austere view of sound promoted by both composers, a view that forms much of the bedrock of modernist listening practices. Perhaps the rhetoric of "sounds themselves" continues to be attractive because it allows us to imagine getting beyond the claustrophobic realm of human purposes. It lures us into thinking that we are delving into the essence of sound, that we are getting past a kind of hearing that instrumentalizes the world according to our designs. In straining to hear sounds themselves, we try not to decipher, to interpret, to extrapolate, or to subjectify, but to revel, presumably, in the sensuous immediacy of vibrations impinging on ears and body. This desire has taken various forms over the years, from Cage's essays to Susan Sontag's polemics against interpretation to Carolyn Abbate's elevation of the drastic over the gnostic.[35] A similar impulse can be witnessed in the rise of speculative realism and object-oriented ontology, both of which try to rise above the correlationism according to which what we can say about the world must be explicitly couched as the product of human modes of perception and cognition.[36] Getting absorbed in "sounds themselves" and shedding the compulsion to interpret, we seem to escape, however fleetingly, the imperatives of what anthropologist and systems theorist Gregory Bateson called "purposive consciousness," that form of linear thinking narrowly focused on human aims.[37]

Does reveling in the sensuous qualities of sounds—especially natural sounds—get us closer to the cyclical, system-oriented perspective Bateson thought we needed to cultivate? If anything, such a listening posture only exacerbates the tendencies that concerned Bateson by isolating individual phenomena from larger contexts and extracting sensuous particulars from their place in larger systems—social, semiotic, economic, material, and so on. If ecology and systems theory have taught us anything, it is to be suspicious of the notion that anything exists wholly in- or for-itself. Treating things as though they do exist in this way, Bateson maintains, is a product of the "distortion" wrought by human consciousness. Even though Rothenberg recommends "dwelling inside an ecology to know the significance of a wayward sound," his separation of beauty from purpose makes it seem as though apprehending the musicality of natural sounds depends on suppressing their ecological significance.[38] To be sure, there is nothing to stop listeners from attempting the "reduced" listening prescribed by the ontology of "sounds themselves."[39] Plenty of modern music, for example, would appear to welcome such an approach. But when reduced listening is transferred from the concert hall to nonhuman habitats, it threatens to reduce the raison d'être of many natural sounds straight out of existence. In *How Forests Think*, Kohn writes that "signs are not exclusively human affairs. All living beings sign. We humans are therefore at home with the multitude of semiotic life."[40] In a similarly ecological spirit, composer David Dunn remarks, "What we hear from other forms of life and the environment they reside in is information that is unique and essential about patterns of relationship in context."[41] If "sounds themselves" are a fiction or, more generously, an asymptote toward which human perception aspires in very limited instances, then human sound making in general evinces the same entwinement of meaning and form exhibited elsewhere in nature, a situation that attests to the connections between, in Adams's words, "the sounds we make and the larger, older world."[42]

Conundrums of Musical Semiosis

In an essay entitled "Parahuman Wagnerism," Gary Tomlinson argues that music is specially poised to reveal these connections, even in cases where it has been understood primarily in terms of symbolic representation (his example is the leitmotivic texture of Wagner's operas). Starting from the premise that semiosis extends "out toward the broadest reaches of the biome," Tomlinson argues that listeners encounter mostly indexical signs in music—signs he describes as "standing near to, gesturing at, pointing to, or indeed causing their objects."[43] Indices, Tomlinson maintains, are minimally referential; they lack the "aboutness" that, in his view, icons and symbols possess.[44] Since many more creatures respond to indexical signs (such as the sounds made by their own and other species) than symbolic ones, Tomlinson classes music among what he calls "informational processes of wide extrahuman dispersion."[45]

Tomlinson's thesis regarding the close connection between musicking and nonhuman semiosis is appealing, especially given music's conspicuous affective impact, its resemblance to phonocoded forms of animal communication (see chapter 5), and the "embodied and palpable" nature of its signs.[46] Indeed, music is psychoactive in a manner that attests to the index-like contiguity between its sonic attributes and the affective states it is capable of inducing—soft, gentle singing for a lullaby, say, or full-bodied, percussive chanting for a war song. But given that Tomlinson reverses the customary way of understanding indices—he refers to them as *causing* their objects rather than being caused *by* them, as the index smoke is caused by its object, fire—it appears that something more than Peircean indexicality is involved in such cases. Moreover, if the listener's responses, rather than musical sounds, assume the role of indices that point to the music that causes them, then one would still need to specify how particular configurations of musical sounds come to wield this causal power.[47]

Tomlinson's approach to musical signification presents further difficulties. For instance, although he builds on Naomi Cumming's Peircean study of music *The Sonic Self*, he passes over Cumming's resituation of the bulk of musical indexicality back into the domain of iconicity, such that a vocal or instrumental "cry" does not signal actual distress but, in Cumming's words, "represent[s] that state 'iconically' at a more abstract level."[48] In other words, a musical cry constitutes a *likeness* of a vocal contour appropriate to the communication of distress. For Cumming, musical expression is typically iconic in this fashion, while the reference of indices is "grounded in their time and place of use."[49] Just as the rotations of a weather vane index the movement of wind, the sounds produced by voices and instruments index the dynamic actions of human bodies. Musical expression, by contrast, unfolds in a virtual world that is distinct, though not entirely separate, from the kinetic arena of sound production. So while Tomlinson, in his book *A Million Years of Music*, argues that tempo and dynamics serve as "energetic indexes" and that vocal contour and rhythmic organization index states of affective and physical arousal, Cumming would understand these features as iconic—as indices once removed or virtualized but not by consequence any less effective.[50] This second-order indexicality, or the creation of virtual worlds out of networks of intramusical relations, probably should not be equated with the indexical signification taking place across the "broadest reaches of the biome."

In "Parahuman Wagnerism," Tomlinson sidesteps the virtual nature of musical indexicality by redirecting attention to the *interpretant*, which, in Peircean terms, refers to the connection an observer makes between a sign and something presumed to be its object. Wagner's music may point to "the dramatic presentation before us," Tomlinson admits, but rather than depicting aspects of the drama, the music instead invites us to construct "psychic states" in such a way that we "*make* the music at every moment."[51] Listening to Wagner, Tomlinson concludes, involves not the passive registration of symbolic meanings but a "subjective activism" that consists of "the making of interpretants that relate indices to their objects."[52] It is precisely because musical expression does *not* depend on direct causality or physical proximity—key aspects of indexical signification—that Tomlinson must take cover under the workings of the interpretant, leaving the reader uncertain about how music actually manages to signify at all. Tomlinson's analytical vocabulary of flat II, subdominant, and V7 chords, along with the larger expressive complexes to which they belong, refers to a level of musical meaning for which he offers little explanation—the level at which musical signs function as elements of a system (in this case, a tonal system). These elements are not themselves indices in any obvious sense. A V7 chord might point to a tonic, but since that tonic need not be realized for V7 to be an effective sign, dominant chords and the like do not display the "contiguity and direct causality of the index."[53] Since he provides no other way to understand the semiotic function of these familiar musical elements, Tomlinson makes it seem as though the expressive import of music is entirely arbitrary rather than deeply enculturated and dependent on the particular musical devices in play.

The situation gets even more complicated in *A Million Years of Music*. Tomlinson's book argues that present-day musicking arose from the "systematization of ancient, indexical gesture-calls," a claim he supports through a speculative reconstruction of the emotive contours and "technosocial" rhythms of early hominin protodiscourse.[54] Yet Tomlinson also envisions an intervening period during which discrete pitches were abstracted from vocal utterances, giving rise to sonic phenomena that were "distanc[ed] . . . from meaning." He explains, "While broad pitch contours continued to convey emotive and even semantic content, the pitches underwent an absolution from signifying."[55] In this stone-age version of "sounds themselves," music becomes the province of pitches that are "barely signs of any kind."[56] For Tomlinson, these not-really-signs serve as the basis for larger expressive units, which then serve as material for an "indexical systematicity" different from the symbolic systematicity of language. But does music really systematize indices? There can be little doubt that hominin protodiscourse was largely indexical in nature, with utterances conveying anything from solace to aggression to a simple "here I am." After discrete pitches began to make their way into such utterances, however, giving rise to fixed patterns of pitches or intervals, the causality and contiguity of indexical reference no longer account for the total semiotic phenomenon. This observation applies to both human music and nonhuman song. Think of the birdsongs I discussed in the previous chapter: although any song points indexically to the bird who is singing, the flexibly varied melodic units or phrases that make up the song are unlikely to be individually indexical of anything. Many birdsongs involve an order of formal play that exceeds the scope of indexicality, even if the robustness and intricacy of the song indicates something like reproductive fitness. Successful songs probably attract admirers because, in some sense only partially available to human ears, they sound good. Discrete pitch is the vehicle for performances that exceed mere functionality—performances in which an excess of signification, rather than abstraction per se (as Tomlinson might have it), serves as the gateway to the aesthetic dimension.[57]

Tomlinson's treatment of meaning, or what he calls "aboutness," also raises questions. While he maintains that aboutness is "all but irrelevant" in musical experience, he also suggests that it "intrudes in human experiences of the world" in a manner that "probably does not extend far beyond our species."[58] That is, we cannot help but think that music is about something, especially in a dramatic setting such as a Wagner opera. We cannot, as Tomlinson puts it, "*not* come to a weighing of a sign/object bond."[59] But what is the Peircean interpretant if not the weighing of a sign/object bond and thus an activity that takes place in all types of semiosis, indexicality included? Restricting aboutness to language-like representation obscures the referential character of other kinds of signs. Kohn writes, for example, that "indices provide information; they tell us something new *about* something not immediately present" (my emphasis).[60] Semiosis in general might be described as an activity through which living beings gather and interpret information about their worlds. The sound of a flag flapping in the wind, for instance, would routinely startle my dog because he perceived that sound as indexically linked to a potential threat in the environment. Stripping sonic information of aboutness is another relic of modern information theory, which, as Terrence Deacon observes, reduces information to physical parameters and casts aside "reference, meaning, and significance."[61] The early twentieth-century biologist Jakob von Uexküll, to whom Tomlinson's biosemiotic perspective is indebted, argued that "the question as to meaning must . . . have priority in all living beings." All life processes, for Uexküll, involve "carriers of meaning" and "meaning factors."[62] Portraying aboutness as an "intrusion" on some a priori, nonreferential semiotic state, a state which allegedly encompasses music as well as the world (or worlds) experienced by nonhumans, not only introduces an artificial division into the fluid process of listening to music but also risks reinforcing invidious distinctions between meaningless nature and meaning-making humans.

My aim here is less to criticize Tomlinson than to demonstrate that music does not submit willingly to Peircean semiotics so much as it demands the extension and refinement of that semiotics. Put another way, music blends aspects of iconicity, indexicality, and symbolism in a fashion that is, if not wholly unique, then uniquely difficult to disentangle.[63] To see this, let us ponder an option Tomlinson rejects; namely, that tones and chords are members of a symbolic (rather than indexical) system by virtue of their systematic interrelationships. The advantage of this perspective is that it makes room for the conventionality, and thus the cultural variability, of musical expression. Cumming calls a symbol "a conventionally stipulated relation (as in most words), requiring knowledge of the convention for its interpretation."[64] However, music generally does not traffic in one-to-one relationships between signs and referents; music does not speak in words, nor are its units of signification as easily parsed as the words of a sentence.[65] Yet music does resemble a symbolic system in that the correspondences between musical materials and what they signify for acculturated listeners are at least partly governed by convention. In this respect, Kohn's remark that symbols "refer to their object indirectly by virtue of the ways in which they relate systematically to other such symbols" does apply to music, if reference is understood as roughly equivalent to expressive significance.[66] Although there is no predetermined relationship between a flat-II chord, or a progression containing that chord, and particular affective or gestural meanings, the expressive qualities of tonal music arise in part from conventionalized differentiations among complexes of musical material and the ability to recognize their resulting connotations. Whatever the alleged indexical meanings of major and minor thirds, for example, the contrasting expressive worlds of major and minor tonalities emerged by way of increasing discriminations within and extensions of the tonal harmonic system.[67] It is in this sense that we can appreciate Adams's remark that human music is a "symbolic and semantic phenomenon in which the relationships *between* sounds mean more than the sounds themselves."[68]

Taking the Peircean approach a step further, one might argue that music's capacity for quasi-symbolic reference rests on a broader basis of iconicity. What this means is that the discrete pitches of music are neither indexical, symbolic, nor meaningless. Rather, discrete pitches are icons—*of each other*. This microlevel iconicity differs from the macrolevel resemblances between melodic contours and expressive vocalization, say, or between musical rhythms and physical gestures. Discrete pitches, short melodic motives, and brief rhythmic patterns are icons that proliferate by way of repetition and variation (again, this applies to some animal songs as well). At this more basic level, icons are, in Kohn's words, "semiotic phenomena, even though they largely lack an indexical component that can be interpreted as pointing to anything other than another instance of the patterns they instantiate."[69] It is the likenesses among pitches, not their indexicality, that allows them to form what Tomlinson calls "arrays," which, in the case of human music, serve as the basis for elaborate musical systems. (What is octave equivalence other than a special kind of iconicity?) Abstracting tones from indexical gesture-calls produces neither meaningless sounds nor deracinated information but a new kind of sign, a sign whose relational possibilities inaugurate the virtual spaces of music. These spaces are broadly symbolic in that their referentiality or expressiveness is, in Kohn's terms, "ultimately the product of a series of highly convoluted systemic relations among icons."[70]

In sum, while music indeed constitutes a complexly embodied and affectively replete mode of communication whose conditions of possibility stretch beyond the boundaries of the human, it does so by way of an arsenal of semiotic strategies that includes the systematic discriminations of the symbolic register. This conclusion, though different in important details from Tomlinson's, shares in his concern to cautiously differentiate between human music and the sonic expressions of other creatures without overstating their points of disconnection. It is not that nonhuman sounds are "sounds themselves" while human sounds are drenched in meaning but that human music, like human language, features a degree of systematic intricacy that appears to be unique in the living world. A more agnostic conclusion would be that even if some nonhuman animals communicate using a mixture of symbolic, indexical, and iconic signs that play out in a virtual realm, we humans would be unlikely to recognize it.

Icons of Absence

The economy of musical signification changes considerably in cases where discrete pitches no longer serve as the primary artistic material. In many contemporary pieces, the systemic, symbolic relationships among tones that Adams singled out for comment cease to be major players in the compositional game. For the Cage of "Experimental Music," dispensing with discrete pitch was the first step toward ushering "nature's manner of operation" into the aesthetic domain.[71] Recording technology would seem to represent another step, in that composers can now collect "wild sounds" and insert them into sound collages whose expressivity bears little resemblance to that of pitched music. A recent example is the sixteen-channel sound installation *Ultrafield* by Norwegian composer Jana Winderen, which was included in the exhibition *Soundings: A Contemporary Score* mounted at New York's Museum of Modern Art in 2013. Winderen's piece featured sounds recorded beneath and around a lake near Oslo—flowing water, melting ice, and the ultrasound emissions of bats and underwater insects transposed into the range of human hearing.[72] While the source material of Winderen's piece may appear to be more natural than the discrete pitches of tonal music, the installation nonetheless relied on a decontextualization not unlike that found in common-practice music, in which the indexical relations between tones and the performer's actions are rendered secondary to intramusical iconicity. That is, *Ultrafield* obscures the originary indexicality of its source sounds to instantiate the virtual space of the installation and the likenesses (and differences) in which it traffics (*that sounds like running water, that sounds like the chirp of an insect*). The bat vocalizations and other ultrasonic sounds are doubly decontextualized by recording and transposition. Winderen's compositional method consists of transforming what were once indices into a nonsystematic arrangement or assemblage of icons.

Although *Ultrafield* could be understood as supporting the commonplace observation that technology expands human perceptual capacities, "allowing us," as the museum placard put it, "to experience sonic realities that are otherwise out of reach," I found myself meditating on a very different prospect as I sat in the dark room housing the installation. What *Ultrafield* indicated to me was that its source sounds, in their broadly semiotic significance, really are out of reach. Listening to transposed versions of ultrasonic sounds only confirmed that I was not really hearing those sounds—that I will never fully understand, for instance, what it is like to be a bat.[73] The powerfully affecting sense of both hearing and not hearing stimulated by Winderen's compositions serves as a reminder that technological access to (or, rather, creation of) "sounds themselves" does not equal access to the embodied and contextual meanings those sounds have for nonhuman others. Adams acknowledges this when he warns that using field recordings in compositions risks reducing living sounds to raw material: "Removing wild sound from its natural context can trivialize and lessen the rich ambiguities inherent in both wild sound and human music."[74] Dunn concurs, arguing that "the sounds of living things are not just a resource for manipulation; they are evidence of mind in nature and are patterns of communication with which we share a common bond and meaning."[75]

Winderen's compositions preserve some of the ambiguity to which Adams alludes by putting listeners in the paradoxical position of hearing what they cannot hear.[76] The attentive listening to wild sounds that informs her music might go some way toward spurring wonder, or even humility, in the face of what does not sound for us in nature. Such an outcome, however, can hardly be guaranteed, however much it echoes familiar hopes for soundscape composition in general. Composers such as Hildegard Westerkamp and John

Levack Drever have sought to distance soundscape composition from musique concrète and acousmatic music aesthetics, both of which place value on divorcing sounds from their sources and composing only with a concern for the "sounds themselves."[77] Westerkamp defines the "essence" of soundscape composition as "the artistic, sonic transmission of meanings about place, time, environment and listening perception," while her colleague Barry Truax affirms that the meanings of environmental sounds used in soundscape compositions are "inescapably contextual."[78] Yet an important component of the meaning of such sounds—their indexical connections to something presently making the sound—is unavoidably lost in the act of recording. Nor can meanings be as easily transmitted by recorded sounds as Westerkamp implies (recall that Shannon's conception of the information contained in a signal had nothing to do with the signal's meaning). She therefore recommends that composers include additional information regarding the place, time, weather conditions, and social context of recording.[79]

Soundscape composers tend to be acutely aware of the problems attendant on their chosen métier. Westerkamp recognizes that although what the microphone affords "feels like access, like closer contact . . . it is in fact a separation, a schizophonic situation."[80] The separation of sound from source only increases when soundscape recordings are commodified and imported into completely different locales, making it difficult to maintain the "ecological thinking" that inspired those recordings in the first place.[81] The danger is that aesthetic absorption in a recorded (or virtualized) soundscape might displace interest in the environment that made it possible, "inadvertently participating," says Westerkamp, "in the place's extinction."[82] She nonetheless remains optimistic, contending that "the 'naked ear' of the microphone can arouse an attentiveness in our listening, which will have a direct influence on how we speak with environmental sounds through our compositions and productions."[83] Drever, citing Truax, is similarly convinced, and he includes among a list of what constitutes a soundscape composition the requirement that "the work enhances our understanding of the world, and its influence carries over into everyday perceptual habits."[84]

If composers are lucky enough to bring about this enhancement, it is not because they have truly overcome the separation between sound and source (and the concomitant loss of contextual, indexical meanings) characteristic of acousmatic sound. Westerkamp reveals as much when she gives credit to "the sound materials themselves" for determining the shape of a composition.[85] What those materials give rise to in the virtual space of the artwork is fundamentally different from the holistic environments to which they originally belonged. "At the point when the ear becomes disconnected from direct contact with the soundscape and suddenly hears the way the microphone 'hears' and the headphones transmit," Westerkamp muses, "at that point the recordist wakes up to a new reality of the soundscape," a reality in which "the sounds are highlighted and the ears are alerted."[86] This is, in short, a virtual reality, one that selects a narrow band of acoustic phenomena from a semiotically multivalent and multisensory environment, captures and strips those phenomena of their in-the-moment indexicality, and transports the resulting sounds into the virtual space of the aesthetic artifact. Schizophonia breeds schizosemiosis; the direct causality of indices gives way to the contingent generation of interpretants that are, essentially, interpretive or hermeneutic. It is up to individual listeners, writes Truax, to "complete the network of meanings" initiated by the work, or, according to Westerkamp's directive, to aspire to "genuine ecological consciousness."[87]

Adams's recasting of composition as "sonic geography" might be understood as an attempt to eliminate some of the bad taste of resource extraction that clings to the technological capture and export of environmental sounds. By composing music that aspires to "be landscape" rather than represent it, Adams continues the long tradition of crafting virtual musical spaces that may or may not profess ties to actual places.[88] This project takes an especially unusual form in *Earth and the Great Weather*, a ten-movement cycle that features a string orchestra, percussion, recorded natural sounds, and spoken litanies.[89] The litanies—in English, Latin, and the indigenous languages of Iñupiaq and Gwich'in—refer to places, seasonal times, directions, and the avian and vegetal occupants of the North American Arctic. Adams has written that the references to places, rather than denoting actual locales, serve as "landmarks for the listener traveling through this country of the ear."[90]

The first movement, titled (like the essay we encountered earlier) "The Place Where You Go to Listen," opens with recorded sounds of wind that coax listeners to imagine being outdoors. The speakers intone such phrases as "You can see a long way," "A point of land which juts into the ocean," "Long, high bluff," and "Place where ice is found," along with many indigenous-language terms left untranslated. Sustained tones in the strings, bowed in such a way as to generate glistening harmonics, rise up from the lower register, their upward trajectory reiterating unconscious associations between registral span and spatial extent. Staggered crescendi and decrescendi spanning the dynamic range *pp* to *ff*, along with an increasing use of tremolo, create an aural environment marked by continuous energetic flux amid relative tonal stasis. The imaginary landscape persists, though it is buffeted by constantly shifting winds.

Listening to the movement, I do not think one is meant to ponder rocky bluffs or thickets of ice in general. Instead, the spoken litanies encourage the work of imagining not just a specific place but also the indexical relations between sounds and things that help define a place. In the essay "Winter Music," Adams writes of music as a "wilderness" one can get "lost in."[91] If being absorbed in music is analogous to being located in a place, then Adams's music aspires to be heard as a series of virtual indexes, as the sonic residue of some imaginary auditory scene. *Earth and the Great Weather* attempts the imaginary restoration of originary indexicality, as if each word and phrase of the litanies pointed to an actual thing and each instrumental or recorded sound arose from a dynamic force or creature in the environment. In other words, the piece construes aesthetic emplacement as a rehearsal for real emplacement. The rhetoric of "Resonance of Place" notwithstanding, Adams's music adheres to a philosophy of music not as the play of sounds themselves, but as a virtual reorientation toward the actual.

How *Waldszenen* Thinks

Let us return to music in which, as Adams maintains, "the relationships *between* sounds mean more than the sounds themselves." In Schumann's *Waldszenen*, we find a complex network of iconic, indexical, and symbolic relationships that creates its own possibilities for how music can "be landscape," in Adams's sense. For instance, the seventh movement, "Vogel als Prophet" ("Bird as Prophet"), imagines a territorial sound "reterritorialized as music," to borrow Elizabeth Grosz's Deleuzian phrase. The virtual song of Schumann's *Vogel* is "not positioned in a definable geographical territory but within a plane of composition in which it summons up primordial fears, desires, and pleasures . . . only to direct them, reterritorialize them, on the plane of music itself."[92] The movement opens with a melody whose expansive tessitura, twittering thirty-second notes, and inscrutable silences suggest the song of nothing human (example 6.1). Rhythmic patterns and melodic shapes proliferate as icons or likenesses of one another: a dotted eighth followed by three thirty-seconds, arpeggiated triads preceded by chromatic appoggiaturas. Following the thread of the melody depends largely on recognizing the repeating rhythmic pattern as well as the intervallic similarities between transpositions, variations, and inversions of the opening four-note figure while not noticing, in Kohn's sense, the differences between them.[93] At this level, musical signs are about each other, although they may also be about affect and gesture. Elements of the music piggyback on one another in a manner resembling what Kohn calls "the iconic propagation of self-organizing thought," where thought is understood not as a special type of

EXAMPLE 6.1. Robert Schumann, *Waldszenen*, op. 82, mvmt. 7, "Vogel als Prophet," mm. 1–27

mental activity but as the dynamic proliferation of signs.[94] It may seem unnecessary to point out that the techniques of variation and elaboration, on which so much music is based, rest on a semiotic foundation of iconicity as the mediation of likeness and difference. But if all semiosis involves meaning, then the iconicity of music effectively refutes the notion of music's meaninglessness, as well as Tomlinson's claim that musical experience offers an "*a priori* to aboutness."[95]

Even as rhythmic and melodic likenesses ricochet through the texture of Schumann's piece, the *unlikeness* of the melody to others in and beyond the cycle serves to spur the further generation of interpretants. A listener unaware of the title would likely wonder what kind of utterance this is meant

EXAMPLE 6.1. (continued)

to be, what it expresses, what it is getting at. The melody invites listeners familiar with the title to hear it as the vocal peregrinations of some fantastical bird—as the indexical utterances of an imaginary creature. Subsequent shifts in thematic material and expressive register indicate that the bird is not the only animal on the scene. Schumann's piece is also about a nonhuman sound and how it is heard by a human listener—heard, that is, as prophetic by some-

EXAMPLE 6.1. (*continued*)

one *in the world of the piece*. In this sense, "Vogel als Prophet" is about what Adams called "attentive listening to wild sounds."

The virtual listener to this virtual birdsong is not just hearing sounds that "simply sound." The first phrase, which lasts from the initial upbeat through the first half of measure 2, sets forth the virtual wilderness of the piece. The call's off-kilter appoggiaturas invest it with a degree of otherness, as if the bird's utterances do not quite fit into the conceptual framework of the human perceiver. This is largely due to the delayed resolution of the appoggiatura's dissonances: although C♯4 and C♯5 proceed to D4 and D5, respectively, it is only with the arrival on D6—and thus with the traversal of a sixteenth rather than a second—that the call truly reaches a place of repose.[96] Further elaboration of the call culminates in a cadence on the relative major B-flat (m. 5), and this harmonic pattern recurs in the restatement of the opening call in D minor and subsequent turn to F major (mm. 5–8). Although it is strange to think of birdsong modulating in this way, the music nonetheless carves out a space for the bird that is neither that of total domestication nor complete incomprehensibility. After a series of phrases that traverse third-related harmonic areas, a pattern whose neatness is skewed by those persistent appoggiaturas, the passage beginning in measure 9 suggests, to my ear, that the responsiveness of the imaginary listener gains a certain momentum along with the vocalizations of the bird. Motivic material deriving from the opening call is passed between right and left hands, while the right-hand melody briefly flirts with a simpler, appoggiatura-free, more human kind of singing (pickup to 11 through 12). It is as if the imagination, or even the voice, of the human listener enters into dialogue with the bird's cries—cries whose capacity to signify, to inspire the further proliferation of signs, displays the future orientation characteristic of prophecy. This dialogue does not last long, however, as the original birdcall returns in measures 16–18.

A sudden shift to the human world follows at the end of measure 18 with the entry of a chorale-like tune in G major. The chorale's metrical displacement, more seen than heard, places it slightly at odds with the sonic space of the birdcall; these two sanctuaries, so to speak, are not entirely contiguous.[97] Laura Tunbridge notes an undercurrent of warning in the melody stemming from its intertextual resonances in Schumann's oeuvre and from the original motto the composer selected for the movement: "Hüte dich, sei wach und munter!" ("Be on your guard, be awake and alert!") from Eichendorff's poem "Zwielicht" ("Twilight").[98] But she also concedes that the chorale comes across as relatively harmless, doing little to upset the course of the movement. While it would be easy to jump to the conclusion that *this* is the prophecy toward which the movement has been heading, the chorale is a little too bland to enjoy any such honor. Could it be that the piece juxtaposes the enigmatic semiosis of natural signs, whose meaning is not fully penetrable by human ears, with the all-too-familiar offerings of conventional religion? As if resigned to its own lack of interest, the chorale (and the virtual world of human music nested within the piece) holds the fictional listener's attention only for a short time. At the marking "Verschiebung" (soft pedal), which also means "displacement," the melody wanders away from its tonal and rhythmic moorings, slows down, and moves into E-flat major. Alternatively, perhaps the chorale, as a feature of human society, actually stays where it is, while the fictional listener's attention drifts back to the wild world of avian song.

These almost entirely disparate musics are linked by one of those tiny motivic connections Schumann relished: the lowly half step, which characterizes both the birdsong's appoggiaturas and the chorale melody's (displaced) first and third beats. At the end of measure 24, the chorale's dotted half step is transposed to D–C♯ so that its inversion C♯–D can launch the return to the birdsong. The last beat of measure 24 is an uncanny moment indeed; one cannot judge where the human music ends and the birdsong begins. This moment could be thought of as a paralinguistic instance of what Friedrich Kittler termed the "minimal signified," or it could be understood in more formal terms as a pattern of utterance shared across species lines.[99] Schumann's half step, a hybrid of human and avian modes of expression, encourages the recognition that iconic similarities are meaningful in their very iconicity, in their ability to reveal the common formal (or morphodynamic) ground traversed by creatures whose physiologies and phylogenies are profoundly different. At

ON NOT LETTING SOUNDS BE THEMSELVES 153

the same time, the appearance of similarity warns us to be on the lookout for difference—which in this case means the possibility that a human half step might be more like a sixteenth to a bird, with much more room to maneuver.

"Vogel als Prophet," in sum, is itself prophetic of a listening attitude that rejects the notion of nature as the domain of "sounds themselves" in favor of appreciating—to the extent that we can—the full-bodied semiosis of nonhuman others. But, the reader may counter, is not all of this happening merely within the virtual environment of a piece of human music? Does Schumann's piece really have anything to do with the songs of real birds? The ending of "Vogel als Prophet" may offer a provisional answer—which might seem strange, considering that measures 25–42 repeat almost exactly what was heard in measures 1–18 (example 6.2; mm. 28–36 not shown). Tunbridge notes that the effect of these final measures is "one of departure rather than conclusion; we leave the bird fluttering around the branches, and then we simply move on to the next scene."[100] The rising flourish and wispy dyad that close the movement merely reiterate what has by now been heard many times over; a definitive sense of tonal closure is deferred to *Waldszenen*'s final movement, "Abschied" ("Departure"). The human visitor's departure from the forest provides the concluding gesture, yet the forest as a site of living thought (to borrow a phrase from Kohn), of the abundant proliferation of

EXAMPLE 6.2. Schumann, *Waldszenen*, op. 82, "Vogel als Prophet," mm. 37–42

signs, continues on as before. Places remain, birds keep calling, flowers keep growing, hunters and wanderers come and go. The inconclusiveness of "Vogel als Prophet" points beyond itself to the open-ended multisensory environments of the outdoors. After giving the piece a good listen, why not leave the space of aesthetic consumption and continue the semiotic adventure elsewhere? At the risk of courting accusations of sentimentality, one could do worse than emulate Kant's "beautiful soul" who trades the gallery and concert hall for field and meadow, eager to discover "a train of thought that he can never fully develop."[101] Nineteenth-century character pieces and twentieth-century sonic geographies may inspire respect for the wider world of living sound, but, as R. Murray Schafer once wrote, "The rest is outside your front door."[102] Let what began with birds—Adams's birds of the north, Schumann's birds of the imagination—end with birds, with the calls of robins, cardinals, chickadees, mourning doves, red-winged blackbirds, blue jays, woodpeckers, flickers, finches, thrushes, sparrows, seagulls, hawks, wrens, and crows. May their trains of thought go on, and on, and on.

Acknowledgments

The research and writing of this manuscript was generously supported by an American Council of Learned Societies Fellowship in 2014–15. I would like to thank all those who read the manuscript at various stages; I am especially grateful for the extraordinary support of Aaron S. Allen, James Currie, Eric Drott, Melina Esse, Lisa Jakelski, and Roger Moseley. In addition, I would like to thank the graduate students and professors in my spring 2016 Aesthetics after Humanism seminar at the Eastman School of Music, where many of the ideas explored here enjoyed lively discussion. Further thanks go to the organizers of symposia and conferences at which I was invited to present my work; these include presentations at Cornell University, Harvard University, the University of Michigan at Ann Arbor, Yale University, and Rutgers University as well as papers and keynotes at the annual meeting of the American Musicological Society, Pittsburgh, 2013; "Music and Nature," Stony Brook University, 2014; "Dissonant Discourses," Vanderbilt University, 2014; "Hearing Landscape Critically," Harvard University, 2015; "Romanticism—Philosophy, Literature, Music," Internationales Zentrum für Philosophie, Bonn, 2015; NUNC! 2, Northwestern University, 2015; the annual meeting of the Music Theory Society of New York State, Hobart and William Smith Colleges, 2017; and the annual meeting of the New York–St. Lawrence Chapter of the American Musicological Society, University of Toronto, 2017. Finally, some of the book's material has previously appeared elsewhere. A portion of chapter 1 was published as "Toward a Post-Humanist Organicism," *Nineteenth-Century Music Review* 14, no. 1 (2017): 93–114; an earlier version of chapter 5 appeared in *Evental Aesthetics* 2, no. 2 (2013); and chapter 4 (with slight alterations) was published in *19th-Century Music* 36, no. 1 (2012): 24–45.

Notes

Full citations for sources cited in notes appear in the bibliography.

Introduction

1. Nils L. Wallin speculates that "one function of music might be its contribution to retaining, restoring, and adjusting an individual's vital character"; see his *Biomusicology*, 481.

2. Higgins, *Music between Us*, 18. Like Higgins, I focus on the enjoyment of music rather than uses of music that harm or disturb by disrupting vital functions. On the latter, see Goodman, *Sonic Warfare*; and Cusick, "'You Are in a Place That Is Out of the World . . .'"

3. See Cox, "Embodying Music."

4. For two complementary perspectives on the virtuality of music, see Langer, *Feeling and Form*; and N. Cumming, *Sonic Self*.

5. Glen A. Love charts a similar path for literary criticism in his *Practical Ecocriticism*.

6. See Varela, Thompson, and Rosch, *Embodied Mind*; Thompson, *Mind in Life*; Bateson, *Steps to an Ecology of Mind*; Luhmann, *Art as a Social System*; Deacon, *Incomplete Nature*; Kohn, *How Forests Think*; Haraway, *When Species Meet*; Marder, *Plant-Thinking*; Grosz, *Chaos, Territory, Art*; and Deleuze and Guattari, *A Thousand Plateaus*.

7. My project is not "vitalist" in the sense conveyed by that term around the turn of the twentieth century. In contrast to vitalism's hypothesis of an independent life force animating matter, this book remains committed to the idea that material processes generate the emergent properties which ultimately constitute life.

8. See Bennett, *Vibrant Matter*. For a critique of empathy in relation to plants, see Marder, "Life of Plants."

9. Appraising the imperative among literary critics to "denaturalize" cultural practices, Rita Felski writes, "It would be hard to overstate the pervasiveness of this antinaturalist rhetoric in contemporary scholarship" (*Limits of Critique*, 70).

10. For examples of these disciplinary crosscurrents, see the special issue of *Cognition* devoted to music (volume 100, no. 1 [2006]); Wallin, *Biomusicology*; Wallin, Merker, and Brown, *Origins of Music*; Marler and Slabbekoorn, *Nature's Music*; and Wheeler, "Lightest Burden."

11. These venues include the journals *Behavioral Ecology*, *Music Perception*, and *Ecomusicology Review*.

12. "Gnostic" refers to the terms of contention in Carolyn Abbate's essay "Music: Drastic or Gnostic?" On "musicking," see Small, *Musicking*. The phrase "biotic arts" comes from Richard O. Prum's *Evolution of Beauty*, 336.

13. See Burnham, *Beethoven Hero*; see also Michael Gallope's discussion, in his *Deep Refrains*, of Schumann's appeal for Deleuze and Guattari, who, as philosophers of the "inhuman," otherwise concentrated their critical energies on modernism.

14. See chapter 3 ("Robert Schumann and Poetic Depth") of my monograph *Metaphors of Depth in German Musical Thought*.

15. Helga de la Motte-Haber's excellent study *Musik und Natur* touches on a range of German-language examples from the nineteenth century; the book's scope ranges from Pythagorean thought to the late twentieth century. For additional studies of interactions between musical discourse and ideas concerning nature, see Hiekel, *Ins Offene?*; Clark and Rehding, *Music Theory and Natural Order*; and Schleuning, *Die Sprache der Natur*.

16. In Aaron S. Allen's terms, my book falls on the "poetic" rather than the "practical" side of ecomusicology. See his "Ecomusicology from Poetic to Practical." See also the many "practical" essays in Allen and Dawe, *Current Directions in Ecomusicology*; see also Sabine Feisst et al., "Music and Ecology."

17. Here I invoke a strand of posthumanism distinct from both the disembodied, information-centric discourses that N. Katherine Hayles terms "posthuman" and "transhumanist" calls for further hybridizations of humans and technology. See Wolfe, *What Is Posthumanism?* and Hayles, *How We Became Posthuman*. See also Haraway's *When Species Meet*, which rejects the term *posthumanism* in favor of a far-reaching notion of "companion species," and Sagan, *Cosmic Apprentice*. For dialogues among posthumanism, ecocriticism, and ecomusicology, see Westling, "Literature, the Environment, and the Question of the Posthuman"; and Edwards, "Critical Theory in Ecomusicology."

18. See, for example, Wolfe, *What Is Posthumanism?*; Morton, *Hyperobjects*; Morton, *Realist Magic*; and Cecchetto, *Humanesis*.

19. On the skewing of materialist music studies toward technology, see Watkins and Esse, "Down with Disembodiment."

20. Lawrence J. Hatab defines *physis* (or *phusis*) as "self-manifesting movement," as the capacity "to grow, to bring forth, to give birth," and as a "dynamic process of actualization." See his "Nietzsche, Nature, and the Affirmation of Life," 34.

21. Hayles, "Cognitive Nonconscious," 808.

22. On the shortcomings of music conceived as a "mind-mind game," see Suzanne Cusick's pivotal essay "Feminist Theory, Music Theory, and the Mind/Body Problem."

23. Dorion Sagan uses the term "hypersex" to refer to the "commingling of organisms that meet, eat, engulf, invade, trade genes, acquire genomes, and sometimes permanently merge" (*Cosmic Apprentice*, 19–20).

24. See the National Human Genome Research Institute's webpage on comparative genomics (https://www.genome.gov). On the need to expand the definition of history to include anthropogenic climate change, see Chakrabarty, "Climate of History."

25. For a helpful survey of the current breadth of such endeavors, see Olivia Bloechl, Melanie Lowe, and Jeffrey Kallberg's introduction to the collection *Rethinking Difference in Music Scholarship*.

26. In a recent essay, Rachel Mundy problematizes adopting "the human" as a category meant to replace differentiations based on the notion of race, because it isolates humans from other species with which they may share cultural and physiological traits. See her "Evolutionary Categories and Musical Style." For an argument that posthumanism needs to engage with racial difference rather than seek shelter under the featureless category of "the human," see Weheliye, *Habeas Viscus*.

27. David Halperin's recent claim that certain nonreproductive manifestations of erotic love take us "away from nature" betrays a typical lack of knowledge regarding the range of sexual (including homosexual) behaviors in the nonhuman natural world. See his essay "What Is Sex For?" For an eye-opening corrective, see chapter 3 ("Sex") of Frans de Waal's *Our Inner Ape*.

28. See Edward O. Wilson's discussion of conservation ethics in *Biophilia*.

29. One can hardly avoid citing Raymond Williams's quip that nature is "perhaps the most complex word in the language" (*Keywords*, 219); see also Arthur O. Lovejoy's comment that nature is "the most pregnant subject for the investigations of philosophical semantics" (*Great Chain of Being*, 14). The meanings of the word *nature* include, among other things, the essence of something (as in "human nature"), the force that directs the course of events both animate and inanimate, and the world of matter as a whole. On the tensions between realist and constructivist views of nature, see Soper, *What Is Nature?* For explorations of how the concept of nature functions in contemporary ecomusicological writings, see Titon, "Nature of Ecomusicology"; and Gautier, "Acoustic Multinaturalism."

30. Malabou, "One Life Only," 438.

31. B. Latour, *We Have Never Been Modern*.

32. On this point, see Grosz, *Nick of Time*, 3–4.

33. See B. Latour, *We Have Never Been Modern*; and Descola, *Beyond Nature and Culture*. On the controversy surrounding the notion of animal culture, see Waal, *Are We Smart Enough to Know How Smart Animals Are?*

34. In addition to Descola's *Beyond Nature and Culture*, see Waal, *Our Inner Ape*; and R. Dunn, *Wild Life of Our Bodies*. Judith Becker defines *culture* as "a supra-individual biological phenomenon, a transgenerational history of ongoing social structural couplings that become embodied in the individual and transmitted into the future through actions" (*Deep Listeners*, 130.)

35. See Currie, "There's No Place." On *physis* and *techne*, see Hatab, "Nietzsche, Nature, and the Affirmation of Life," 35; on the overlap and divergence between the ancient notion of *physis* and the contemporary concept of nature, see Holmes, "Before Nature?"

36. For this reason, I object to using the products of techne to metaphorize or conceptualize the products of *physis* (as when organisms or life processes or individual organs, such as the brain or heart, are likened to machines), at least when no provision is made for some remainder that falls outside of such comparisons (see Watkins and Esse, "Down with Disembodiment"). As for isolating *physis* from techne, one might argue that even this becomes difficult in an epoch defined by anthropogenic climate change.

37. See Abrams, *Mirror and Lamp*.

38. Nietzsche, *Beyond Good and Evil*, 3.

39. Ibid., 17–18.

40. Nietzsche, *Genealogy of Morals*, 137–38.

41. Nietzsche, *Beyond Good and Evil*, 8.

42. Hatab, "Nietzsche, Nature, and the Affirmation of Life," 35.

43. See Uexküll, *Foray into the Worlds of Animals and Humans*.

44. For a recent contribution to the debate over realism and constructivism, see Meillassoux, *After Finitude*.

45. See Soper, *What Is Nature?* 4. In a similar spirit, Holmes writes, "However much we may want to throw out 'nature' altogether or sideline it in the practice of ecocriticism, it still remains deeply embedded in how we organize our own thinking about the nonhuman world, haunting forms of scientific inquiry and the epistemic models of literature alike" ("Before Nature?" xii).

46. Nietzsche, *Beyond Good and Evil*, 16.

47. Lemm, introduction to *Nietzsche and the Becoming of Life*, 2.

48. Friedrich Nietzsche, "Schopenhauer as Educator," 194.

49. See Bennett, *Vibrant Matter*, and Grusin, *Nonhuman Turn*.

50. Cage, "Experimental Music," 10.

51. Adams, "Resonance of Place," 17.

Chapter One

1. Or a rhizome: "musical form, right down to its ruptures and proliferations, is comparable to a weed, a rhizome." Deleuze and Guattari, *A Thousand Plateaus*, 12.

2. Adorno, "Vers une musique informelle," 306. On Adorno's earlier critique of organicism, see Buch, "Adorno's Schubert."

3. Hoffmann, *E. T. A. Hoffmann's Musical Writings*, 94.

4. Anonymous translation from the 1992 reissue of the 1966 Bayreuth recording of the opera (Philips 434 425-2).

5. Thaler, *Organische Form in der Musiktheorie*; Schmidt, *Organische Form in der Musik*. The literature on organicism is substantial: for general accounts of organicism's history, including its ancient sources, see G. Rousseau, ed., *Organic Form*; and Abrams, *Mirror and Lamp*. Studies and critiques of music-related organicist discourse include Solie, "Living Work"; Kerman, "How We Got Into Analysis"; Levy, "Covert and Casual Values"; Daverio, *Nineteenth-Century Music*; Clark and Rehding, *Music Theory and Natural Order*; and Neubauer, "Organicism and Music Theory." On the persistence of organicism among post-Romantic composers, see Neff, "Schoenberg and Goethe"; Grimley, "Organicism, Form, and Structural Decay"; and Tarasti, "Metaphors of Nature and Organicism." On the contentious status of organicism in Heinrich Schenker's thought, see Pastille, "Heinrich Schenker, Anti-Organicist"; Cherlin, "Hauptmann and Schenker"; Korsyn, "Schenker's Organicism Reexamined"; and Duerksen, "Schenker's Organicism Revisited."

6. See, for example, Hegel, *Elements of the Philosophy of Right*, sec. 269. For an overview of Hegel's organicism, see Beiser, *Hegel*, chap. 4 ("The Organic Worldview"). Alexander Rehding discusses analogies between musical forms and state organization in August Halm's thought; see his "August Halm's Two Cultures." Finally, Anne Harrington's book *Reenchanted Science* challenges the assumption that early twentieth-century biological organicism is inherently conservative or nationalist.

7. Morton, *Ecological Thought*, 84. See also Chua, *Absolute Music*; and Cusick, "Musicology, Gender, and Feminism."

8. Thaler, *Organische Form*, 130.

9. See Pollan, "Intelligent Plant"; and Thompson, *Mind in Life*. While eighteenth-century commentators sometimes compared the creativity of artists to "vegetable genius," such plant-

like unconsciousness was less valued in itself than treated as a stepping-stone to higher spiritual achievements. See Abrams, *Mirror and Lamp*, chap. 8.

10. Marder, *Plant-Thinking*, 10.

11. Pollan, "Intelligent Plant," 105.

12. Mancuso and Pollan quoted in ibid., 105.

13. Clark and Rehding write, "While it is part and parcel of any idea of nature to rely on the assumption of an essence independent from historical or cultural context, the images invoked in support of this idea will necessarily reflect the culture and historical age in which they arose" (introduction to *Music Theory and Natural Order*, 2). This seems to me unobjectionable; and yet it does little to reduce the difficulty of determining what we are doing—and what we are talking about—when we make truth claims about the natural world. On the problem of correlationism ("the idea according to which we only ever have access to the correlation between thinking and being, and never to either term considered apart from the other") as it pertains to scientific knowledge, see Meillassoux, *After Finitude* (quotation at 5).

14. Adorno, "Vers une musique informelle," 306.

15. Kant, *Critique of Judgement*, 217.

16. Ibid., 218.

17. Ibid., 219–20 (emphasis in original).

18. Ibid., 218.

19. Goethe, *Scientific Studies*, 58.

20. Ibid., 64.

21. Haraway, *When Species Meet*, 31.

22. Hegel, *Philosophy of Nature*, 377.

23. Ibid., 303 (emphasis in original).

24. Ibid., 356.

25. Michael Broyles identifies dynamism, wholeness, and teleology as the three primary concerns of organicism; see his "Organic Form and the Binary Repeat." Daverio portrays the "worn out" version of organicism as unrealistically committed to "the subordination of the individual parts to a larger entity, the perfect concord between detail and overriding design, [and] the natural unfolding of a form from within as if from a germinal cell" (*Nineteenth-Century Music*, 184–85).

26. For a survey of these developments, see Richards, *Romantic Conception of Life*.

27. Adorno, "Vers une musique informelle," 307. Advances in the study of developing embryos inspired the rise of biological organicism—a nondualistic alternative to vitalist and mechanistic theories—in the early twentieth century. See Harrington, *Reenchanted Science*; and Haraway, *Crystals, Fabrics, and Fields*.

28. Hoffmann, "Recension," 634.

29. For more on the contradictory nature of Hoffmann's image, see the first chapter of my book *Metaphors of Depth*.

30. Hanslick, *On the Musically Beautiful*, 81. Translation altered; for the original, see Strauß, *Eduard Hanslick*, 167–68.

31. Marx, *Musical Form in the Age of Beethoven*, 187 (the remark is from Marx's 1859 *Ludwig van Beethoven: Leben und Schaffen*). Pastille points to Schenker's similar rejection of causality, and thus of organic necessity, in his early writings on music; see "Heinrich Schenker, Anti-Organicist."

32. Marder, *Plant-Thinking*, 108.

33. Ibid., 97–105.

34. And a point that has been long in the making: the biologist J. S. Haldane, for example, cautioned against vitalist approaches in which "organisms are regarded apart from their environment." See his *Philosophy of a Biologist*, 53.

35. Deacon, *Incomplete Nature*, 302. See also Elizabeth Grosz's summary of Raymond Ruyer's philosophy, which proposes that both organisms and inorganic entities such as atoms display a finalist "directionality" consisting of "not a final state or stage, an end or telos but a trajectory of continuing elaboration/transformation . . . a becoming" (*Incorporeal* 212).

36. Elaine P. Miller makes a similar point in her *Vegetative Soul*, 10.

37. Deacon, *Incomplete Nature*, 273.

38. Thompson, *Mind in Life*, 38. See also Marder, who writes, "If incompletion means open-endedness, then vegetal growth fully satisfies this . . . in that it knows neither an inherent end, nor a limit, nor a sense of measure and moderation" (*Plant-Thinking*, 24).

39. Michaelis, "Ein Versuch," 683.

40. Ibid., 676.

41. Motte-Haber, *Musik und Natur*, 156.

42. Ibid., 675.

43. See, for example, Thompson's reflections on Husserl in chapter 11 of *Mind in Life*.

44. Tomlinson refers to this as "an effect akin to a phase transition"; see his *A Million Years of Music*, 168.

45. Bregman, *Auditory Scene Analysis*, 528. Compare Adorno's comment that "mere tones" arrived at by "physicalistic procedures" are unable to "become art," which raises the question whether what he considered to be aesthetic semblance is related, at least in music, to the phenomenon of emergence (Adorno, *Aesthetic Theory*, 103). See also Susanne Langer, who claims in *Feeling and Form* that a "work of art . . . is more than an 'arrangement' of given things. . . . Something emerges from the arrangement of tones or colors, which was not there before" (40).

46. Michaelis, "Ein Versuch," 675 ("Einbildungskraft und innere Empfänglichkeit").

47. See Camazine et al., *Self-Organization in Biological Systems*.

48. On self-organization in relation to music, see Tomlinson, *A Million Years of Music*; Jean-Julien Aucouturier, "Hypothesis of Self-Organization"; and Kollias, "Self-Organising Work of Music," 192–99. In a somewhat different vein, Edward W. Large has mathematically modeled human perception of meter using a pattern-forming, nonlinear dynamical system, which performed favorably in comparison to human subjects. See his "On Synchronizing Movement to Music."

49. Two precedents for such an approach are DeLanda, *New Philosophy of Society*; and Kohn, *How Forests Think*.

50. Luhmann, "Self-Organization and Autopoiesis," 147–48.

51. Narmour, *Analysis and Cognition of Melodic Complexity*, 8.

52. Meyer, *Style and Music*, 3. In contrast to my approach, Meyer sharply distinguishes the patterning involved in style from patterns in the natural world because humans alone (he claims) make choices. Yet Meyer immediately qualifies this claim in a manner that casts considerable doubt on the distinction to begin with (4–6).

53. See Camazine et al., *Self-Organization in Biological Systems*, 12. Obviously, not every score serves as a blueprint for sounds, patterned or otherwise (think of John Cage's 4′33″). Kollias's "The Self-Organising Work of Music" argues that performances consist of self-organizing interactions between performers, scores, and acoustic spaces, out of whose "local interactions" the work of music emerges.

54. Wallin, *Biomusicology*, 18.

55. "The sonic objects of music derive only in a very indirect way from the real instruments that are playing" (Bregman, *Auditory Scene Analysis*, 459). Compare Wallin's notion that music generates "virtual time" (*Biomusicology*, 16).

56. Kohn, *How Forests Think*, 167. It is important to note that "lower-order" in this context means not of lesser value but less dependent on additional enabling conditions.

57. Bonds, *Absolute Music*, 103, 108–9. On the shifting fortunes of the notion of autonomy in musicology, see Whittall, "Autonomy/Heteronomy."

58. I am leaving the distinction between organicism as creative strategy and as critical approach intentionally vague here, because, in keeping with a systems/environment perspective, I do not believe it resides exclusively in either.

59. Adorno, *Aesthetic Theory*, 114; see also Susanne Langer's comments on "the semblance of spontaneous movement" in music (*Feeling and Form*, 130).

60. Wallin, *Biomusicology*, 2.

61. Kant, *Critique of Judgement*, 220, 217.

62. Bonds, *Absolute Music*, 105.

63. Burnham, *Beethoven Hero*, 118. This attention to process, however, was frequently subordinated to the competing understanding of the "Beethovenian *telos*" as "not just an end but an end accomplished" (122). Such finality, to reiterate, spells death to an organism.

64. Thompson, *Mind in Life*, 44.

65. Ibid., 39.

66. Hanslick, *On the Musically Beautiful*, 13; Thompson, *Mind in Life*, 44.

67. Thompson, *Mind in Life*, 45. Compare Wallin's definition of music as an "open system of evolving structures growing into sounding artifacts" (*Biomusicology*, 16).

68. James Grier provides a thoughtful reflection on the relationship between texts and works in *The Critical Editing of Music*: "The piece . . . resides equally in the score and in the performing conventions that govern its interpretation at any particular historical moment. . . . The identity of the work, then, varies with the conventions under which the score is understood" (22); "the work exists in a potentially infinite number of states, whether in writing (the score) or in sound (performance); the text is one of those states" (23).

69. See, for example, Wolfe, *What Is Posthumanism?*; Rampley, "Art as a Social System"; and Moeller, *Radical Luhmann*.

70. Moeller, *Radical Luhmann*, 127.

71. Luhmann, *Social Systems*, 40.

72. On assemblage theory, see DeLanda, *New Philosophy of Society*.

73. See Maturana and Varela, *Autopoiesis and Cognition*.

74. Even biological applications of this term, let alone sociological ones, are not universally accepted; see Haraway, *When Species Meet*, 32–33.

75. See Rampley, "Art as a Social System," 24.

76. Habermas, *Philosophical Discourse of Modernity*, 372.

77. Judith Becker has adopted Maturana and Varela's concept of "structural coupling" to theorize phenomena such as rhythmic entrainment, trancing, and music's reinforcement of social bonds (*Deep Listeners*, 119–22). Wallin describes music as relying on a "feedback loop" composed of "three sub-systems": the auditory system, the sounding structures of musical artifacts, and the space-time environment (*Biomusicology*, 16).

78. Luhmann, *Art as a Social System*, 51.

79. Harro Müller, "Luhmann's Systems Theory."

80. For a contemporary attempt to apply concepts drawn from evolutionary theory to musicological analyses of style change, see J. Cumming, *Motet in Age of Du Fay*.

81. Müller, "Luhmann's Systems Theory," 47.

82. Luhmann, *Social Systems*, 37. Burnham has argued that the autonomy of music is precisely what makes criticism—one form of broader contact with the world—possible. See his essay "How Music Matters."

83. See Webster, "Eighteenth Century as a Music-Historical Period?" For a more extended application of Luhmann's social systems theory to music, see Janz, *Zur Genealogie der musikalischen Moderne*.

84. See Gjerdingen, *Music in the Galant Style*, 439; and Rosen, *Classical Style*, 46–47.

85. Webster, "Eighteenth Century," 58.

86. These terms are roughly equivalent to "introversive" and "extroversive" semiosis, as employed by such authors as Roman Jakobson, Kofi Agawu, and Richard Taruskin. See Taruskin's elegant elucidation of the "Joke" quartet in the *Oxford History of Western Music*, 542–54.

87. Chua, *Absolute Music*, 210.

88. In addition to Chua and Webster, see Wheelock, *Haydn's Ingenious Jesting*, 6.

89. Chua, *Absolute Music*, 209.

90. Moeller, *Radical Luhmann*, 22.

91. Thompson, *Mind in Life*, 38–39. A more conventional approach has been, in Burnham's words, to consider Beethoven's heroic works as "closed systems, self-generating, self-sustaining, and self-consuming" (*Beethoven Hero*, 118).

92. Matthias Tischer suggests that intertextuality serves as a corrective to organicism's focus on autonomy and integration; I am suggesting, however, that the two perspectives are not incompatible. See his "Zitat—*Musik über Musik*—Intertextualität."

93. Luhmann, *Art as a Social System*, 204.

94. Wagner, *Opera and Drama*, 339 (translation slightly altered); for the original, see Wagner, *Gesammelte Schriften und Dichtungen*, 194.

95. Luhmann, *Art as a Social System*, 53.

96. Adorno, "Vers une musique informelle," 309.

97. Richard Wagner, "On Franz Liszt's Symphonic Poems," quoted in Dahlhaus, *Idea of Absolute Music*, 26.

98. Langer, *Feeling and Form*, 81.

99. Hanslick, *On the Musically Beautiful*, 30.

100. See, for example, Derrida, "White Mythology."

101. Luhmann, "Self-Organization and Autopoiesis," 153.

102. Adorno, *Aesthetic Theory*, 107.

103. Ibid., 78.

104. Ibid., 109.

105. Webster, *Haydn's "Farewell" Symphony*, 356.

106. Luhmann, *Art as a Social System*, 72.

107. Dahlhaus, *Foundations of Music History*, 14.

108. Ibid., 17.

109. See White, *Metahistory*.

110. Treitler, "Historiography of Music," 362.

111. DeLanda, *New Philosophy of Society*, 10.

112. Ibid., 11–12.

113. Kant, *Critique of Judgement*, 80. On the notion of economy, see Levy, "Covert and Casual Values."

114. Daverio, *Nineteenth-Century Music*, 184; Neubauer, "Organicism and Music Theory," 33.

115. Krieger, *Reopening of Closure*, 35.

116. Absolute distinctions between organic and mechanical or technological registers are, like those between nature and culture, difficult to sustain. However, there can be heuristic value in distinguishing (where possible) between what humans have and have not created, even though humans are not the designers of their own designing. For further reflection on this topic, see Watkins and Esse, "Down with Disembodiment."

117. On these topics, see Schäfer and Sedlmeier, "Does the Body Move the Soul?" Phillips-Silver, "On the Meaning of Movement"; and Cox, "Embodying Music." This literature will be examined further in chapter 5.

118. Kawohl, "Organismusmetaphern," 164.

119. Marder, *Plant-Thinking*, 84.

Chapter Two

1. Hanslick, *On the Musically Beautiful*, 11–12 (in this chapter hereafter cited in text as *OMB*). I generally reproduce Geoffrey Payzant's translations here; noted changes have been made in consultation with Strauß, *Eduard Hanslick*.

2. Ibid., 82. On the debts Hanslick's formalism owed to other music critics, and on his ambivalence regarding the matter of music's spiritual import, see Bonds, *Absolute Music*, chap. 9. For a recent reassessment of Hanslick's legacy, see the essays collected in Grimes, Donovan, and Marx, *Rethinking Hanslick*. Nick Zangwill develops a robustly Hanslickian aesthetics of music in *Music and Aesthetic Reality*.

3. One classic account of this transformation is Edward Rothstein's "Americanization of Heinrich Schenker." For a critique of the "purely musical," see McClary, *Conventional Wisdom*.

4. Kerman, *Contemplating Music*.

5. See McClary, *Feminine Endings*; and Kramer, *Classical Music and Postmodern Knowledge*.

6. Taruskin, "Et in Arcadia Ego," 2.

7. For representatives of these views, see McCreless et al., "Contemporary Music Theory and the New Musicology."

8. Among those in the former category, Nicholas Cook stands out for the way he deftly mediates between music's technical "workings" and its ability to convey meaning in tandem with the mutable conditions of reception. See his "Theorizing Musical Meaning."

9. See Abbate, *Unsung Voices*.

10. See Abbate, "Music—Drastic or Gnostic?"

11. Scherzinger, "Return of the Aesthetic," 253.

12. Ibid., 264, 272.

13. Currie, "Music After All," 151.

14. Ibid., 157.

15. Ibid., 180.

16. See, for example, Scherzinger's "Anton Webern and the Concept of Symmetrical Inversion."

17. This is a problem that has long troubled the discipline of aesthetics, which tends to take art beauty rather than natural beauty as its primary focus. See Carlson and Berleant, "Introduc-

tion: The Aesthetics of Nature"; and Hepburn, "Contemporary Aesthetics." For a reconstruction of the aesthetics of natural beauty in conjunction with contemporary ecology, see Böhme, *Für eine ökologische Naturästhetik*.

18. Grosz, *Nick of Time*, 95.

19. Kohn, *How Forests Think*, 158. On the relationship between formal constraints and musical style, see chapter 1 of Leonard Meyer's *Style and Music*. More recently, Gary Tomlinson has alluded to formalism's "deep history" in the "emergence of cognitive abstractions" in hominin evolution, but his account remains resolutely anthropocentric, even if the *anthropos* in question encompasses more than just modern *homo sapiens*. See Tomlinson, *A Million Years of Music*, 288–89.

20. Anthony Pryer draws attention to the shortcomings of Hanslick's project, writing that "what these methods [of scientific investigation] eventually produced was a list of ingredients of music that seem to be hopelessly disconnected and diverse" (Pryer, "Hanslick, Legal Processes, and Scientific Methodologies," 52). For more on the professional and ideological context of Hanslick's treatise, see Karnes, *Music, Criticism, and the Challenge of History*.

21. Ibid., 30. In his essay "Aesthetic Amputations," Mark Evan Bonds points out that although Hanslick attempted to replace the subjectivism of Kant's and Hegel's aesthetics with scientific objectivity, "many of Hanslick's most distinctive views of the nature of music are adumbrated in the aesthetics of idealism" (8).

22. Kant, *Critique of Judgement*, 174 (in this chapter hereafter cited as *CJ*). I have used this J. H. Bernard translation in consultation with the German text of *Kritik der Urteilskraft*.

23. Bonds describes this move as the "partitioning of essence from effect" (*Absolute Music*, 141). On this partitioning and its consequences for Hanslick's formalism more broadly, see Schmidt, "Arabeske."

24. See, for example, Eva Schaper's discussion of the problem in "Free and Dependent Beauty."

25. See Hoffmann, "E. T. A. Hoffmann: '[Review: Beethoven's Symphony no. 5].'"

26. For an analysis of this idea along with other aspects of Kant's aesthetics, see Budd, *Aesthetic Appreciation of Nature*.

27. Translation altered. The original reads "Solche Zergliederung macht ein Gerippe aus blühendem Körper" (Strauß, *Eduard Hanslick*, 50).

28. See Strauß, *Eduard Hanslick*, 49. On the propensity of analysts to state reasons for musical behavior conceived as intrinsically agential, see Maus, "Music as Drama," 64.

29. On the characteristics of natural purposes, see *CJ*, 216–22. Peter McLaughlin offers an extended commentary in *Kant's Critique of Teleology in Biological Explanation*.

30. On this point, see Miller, *Vegetative Soul*, 22.

31. Thompson also remains aware of the fact that anything we ascribe to nature is still conceived from a human, and therefore limited, perspective. See his *Mind in Life*, 138.

32. See Camazine et al., *Self-Organization in Biological Systems*.

33. Eduardo Kohn cautions against assuming that hierarchies in the nonhuman world carry moral significance (*How Forests Think*, 168). Fred Lerdahl and Ray Jackendoff define a hierarchical structure as "an organization composed of discrete elements or regions related in such a way that one element or region subsumes or contains other elements or regions" (*Generative Theory of Tonal Music*, 13). For an idiosyncratic study of self-similarity in music, see Madden, *Fractals in Music*.

34. Thompson cites biologist Enrico Coen's speculations on human creativity: "We can have

NOTES TO PAGES 52–56

an intuitive notion of someone painting a picture or composing a poem without following a defined plan. Yet the outcomes of such creative processes—the painting or the poem—are not random but highly structured . . . human creativity comes much nearer to the process of development than the notion of manufacture according to a set of instructions, or the running of a computer program" (*Mind in Life*, 180).

35. Mabey, *Cabaret of Plants*. Mabey's claims recall M. H. Abrams's discussion of eighteenth- and nineteenth-century theories of "vegetable genius" in *The Mirror and the Lamp*.

36. Thompson, *Mind in Life*, 87.

37. These remarks should not be taken as implying that music possessing these qualities is aesthetically superior to music that does not. Andrew Mead warns against using putatively natural principles to dismiss particular styles of music—especially serial music—in "Cultivating an Air."

38. Larson, "Musical Forces and Melodic Patterns," 56 (all caps omitted). Larson considers the perception of musical forces to arise out of the combination of two cognitive metaphors: "MUSIC IS MOTION" and "MUSIC IS PURPOSEFUL." Similarly, Maus argues that "the notion of action is crucial in understanding" tonal music such as Beethoven's ("Music as Drama," 65); the action he has in mind is distinctly agential. Compare Seth Monahan's exploration of "interpretations that regard musical objects or gestures as volitional, as purposive, in such a way that is indicative of psychological states," in "Action and Agency Revisited," 324–25.

39. See, for example, Meyer, *Music, the Arts, and Ideas*. Resisting the teleological convictions documented in *Beethoven Hero*, Scott Burnham explores how certain works by Haydn and Beethoven question "the idea of return as resolution and the closely related ideas of teleological process and unequivocal completion"; see his "Second Nature of Sonata Form." For a reconsideration of the role of teleology in repertories other than common-practice European music, see Fink, "Goal-Directed Soul?"

40. On phenomenological experiences of groove, see Danielsen, *Presence and Pleasure*.

41. Schenker, *Free Composition*, 9; see also Schenker, *Neue Musikalische Theorien*, 27.

42. Rothfarb, *Ernst Kurth*, 11.

43. Zuckerkandl, *Sound and Symbol*, 368.

44. Langer, *Feeling and Form*, 126.

45. Ibid., 31.

46. Ibid., 99, 32.

47. See Taruskin, *Text and Act*, chap. 4.

48. Cox, "Embodying Music."

49. See Daverio, *Nineteenth-Century Music*, 24–27, and Schmidt, "Arabeske."

50. Discussing Hegel's association of vegetal life with "regularity and symmetry," Michael Marder counters that "the plant enlivens, sets in motion, and liberates the geometrical arrangement from its own rigid confines through a unique exploration of the straight line that it embodies" (*Plant-Thinking*, 125, 127).

51. Langer, *Feeling and Form*, 126–27.

52. See ibid., 61–68, for Langer's discussion of the semblance of movement created by decorative patterns.

53. Marder, *Plant-Thinking*, 29.

54. Fred Everett Maus explores the erotic dimension of Hanslick's images of music's "body" and "circulatory system" in "Hanslick's Animism."

55. Thompson, *Mind in Life*, 152.

56. Other examples of the title include Marin Marais, "L'arabesque," from the fourth book of *Pièces* for viol and continuo (1717), and Claude Debussy, "Deux Arabesques" for piano (1891).

57. See chapter 4 for further discussion of Schumann's compositions for amateurs. For brief overviews of the piece, see Edler, "*Arabeske* op. 18, *Blumenstück* op. 19"; and Draheim, "Arabeske für Klavier op. 18; Blumenstück für Klavier op. 19." In *Nineteenth-Century Music and the German Romantic Ideology*, Daverio reserves the concept of arabesque (which he derives from Friedrich Schlegel's aesthetics) for his discussion of Schumann's *Fantasie*, op. 17.

58. Andreas Weber, "Cognition as Expression," 160.

59. Langer, *Feeling and Form*, 128.

60. Deleuze and Guattari, *A Thousand Plateaus*, 270.

61. Marder, *Plant-Thinking*, 12.

62. See Daverio, *Nineteenth-Century Music*, 26–27.

63. Marder, *Plant-Thinking*, 110.

64. This is not to say that motivic relations are completely absent: a motivic echo of the opening melody is located in the middle of the new phrase rather than at the beginning.

65. Kant, *Critique of Judgement*, 79 (translation slightly altered).

66. I thank Annette Richards for this suggestion.

67. Edler, "*Arabeske* op. 18, *Blumenstück* op. 19," 251.

68. See Thompson on autonomous systems that "determine a domain of possible interactions with the environment" (*Mind in Life*, 44).

69. Weber, "Cognition as Expression," 160–61.

70. Scarry, "Afterword: An Interview with Elaine Scarry," 272.

71. Weber, "Cognition as Expression," 61.

72. Bonds argues that Hanslick's recognition of the transitory nature of musical styles does not threaten his essentialist concept of musical beauty (*Absolute Music*, 176–77). Yet if beauty is so easily shed from actual works of music, it is hard to see what is truly essential about it or how it can be securely tethered to tonal forms.

73. Hanslick's phrase is "Die Zeit ist auch ein Geist und schafft ihren Körper" (Strauß, *Eduard Hanslick: Vom Musikalisch-Schönen*, 95).

Chapter Three

1. See Gagliano, "Green Symphonies."

2. Gagliano, "Singing Plants at Damanhur."

3. Marder and Gagliano, "Michael Marder and Monica Gagliano: How Do Plants Sound?" Compare Sophia Roosth's argument that transduction, or the "conversion of a signal from one medium to another," supports implicit claims regarding the agency of nonhuman others while also functioning as a mode of technological "imperialism"; see her "Screaming Yeast," 338, 335.

4. Schopenhauer, *World as Will and Representation*, 1:256 (hereafter cited as *WWR*; some spellings have been Americanized). See also Schopenhauer, *Die Welt als Wille und Vorstellung*, vol. 1.

5. Ferrara, "Schopenhauer on Music," 186; Goehr, "Schopenhauer and the Musicians," 205.

6. Lovejoy, *Great Chain of Being*.

7. Bennett, *Vibrant Matter*, 60.

8. Refuting in advance the Gaia hypothesis, Schopenhauer writes, "The talk, so fashionable in our day, of the life of the inorganic, and even of the globe . . . is absolutely inadmissible. The

predicate life belongs only to what is organic" (*WWR*, 2:296. See also Schopenhauer, *Die Welt als Wille und Vorstellung*, vol. 2). More recent commentators have noted that, as Timothy Morton puts it, "drawing distinctions between life and nonlife is strictly impossible, yet unavoidable" ("The Mesh," 24). See also Helmreich, *Sounding the Limits of Life*.

9. Safranski, *Schopenhauer and the Wild Years*, 3.

10. Schopenhauer's "radically different" is "von Grund aus verschieden" (*Die Welt als Wille und Vorstellung*, 2:202). Christopher Janaway notes that Schopenhauer placed human action "on a continuum with animal behavior and other organic processes"; see his "Will and Nature," 396.

11. Barbara Hannan portrays him as such in her *Riddle of the World*. She also notes that she strongly identifies with Schopenhauer's philosophy despite his offensive commentary on women, which seems to have originated in a troubled relationship with his mother.

12. Many recent exponents of materialism question the reductive tendencies of eliminative materialism, which Terrence Deacon describes as an approach in which all "ententional" phenomena, meaning phenomena related to nonpresent ends, are assumed to be explicable by recourse to physical mechanisms. See his *Incomplete Nature*, chap. 3; see also Coole and Frost, *New Materialisms*.

13. See Schopenhauer, *On the Will in Nature*, 42.

14. Marder, "Life of Plants."

15. This phrase is inspired by Eduardo Kohn's *How Forests Think*.

16. Atwell, *Schopenhauer on the Character of the World*, 81.

17. Cartwright, *Schopenhauer: A Biography*, 142–44.

18. Hannan writes that Schopenhauer's claim that will is the in-itself of all phenomena "does not follow logically from the premises. It is at best a suggestion based on an analogy" (*Riddle of the World*, 20).

19. On Schopenhauer and emergence, see Hannan, *Riddle of the World*, 8, 14, and 68.

20. Compare Deacon's effort in *Incomplete Nature* to distinguish thermodynamics, morphodynamics (self-organizing inorganic processes), and teleodynamics (self-organizing organic processes) from one another rather than reduce more complex forms of organization to less complex ones.

21. Bennett writes, "My 'own' body is material, and yet this vital materiality is not fully or exclusively human. My flesh is populated and constituted by different swarms of foreigners. . . . The *its* outnumber the *mes*. In a world of vibrant matter, it is thus not enough to say that we are 'embodied.' We are, rather, *an array of bodies*, many different kinds of them in a nested set of microbiomes" (*Vibrant Matter*, 112–13). Compare *Scientific American*'s recent special feature "The Microbiome," whose editor David Grogan writes, "Leading scientists . . . now think of humans not as self-sufficient organisms but as complex ecosystems colonized by numerous collaborating and competing microbial species" (Grogan, "Microbiome," S2).

22. Schopenhauer, *On the Will in Nature*, 42–43.

23. Hoffmann, *E. T. A. Hoffmann's Musical Writings*, 94. Incidentally, the biologist Jakob von Uexküll advanced a "composition theory of nature" in which the components of an ecosystem resemble the voices of counterpoint. See his *Foray Into the Worlds of Animals and Humans*.

24. Magee, *Philosophy of Schopenhauer*, 154.

25. Hoffmann, *E. T. A. Hoffmann's Musical Writings*, 163.

26. Ibid., 164–65. For a contemporary consideration of the Romantic trope of the "language of nature," see Kate Rigby, "Earth's Poesy."

27. Atzert, "Musik und Freiheit vom Willen?"

28. Nietzsche, *Will to Power*, 430.

29. Schopenhauer, *On the Will in Nature*, 43.

30. Music psychologist Jessica Phillips-Silver and her colleagues describe rhythmic entrainment as involving the "integration of information across multiple sensory modalities," likely including auditory, motor, and vestibular systems; see Phillips-Silver, "Ecology of Entrainment," 4.

31. See Cox, "Embodying Music"; see also Kathleen Marie Higgins's broad overview of research on music, emotion, and the body in her *Music between Us*.

32. See, for example, Becker, *Deep Listeners*.

33. "The assertion that all the movements of our body, even the vegetative and organic ones, proceed from the *will*, by no means implies that they are voluntary . . . the movement of the interior economy of the organism, like that of plants, is guided by *stimuli*" (Schopenhauer, *On the Will in Nature*, 38).

34. Hannan, *Riddle of the World*, 113–14.

35. Ibid., 114–15.

36. Brian Kane describes this passage as revealing the "bodily techniques" involved in an epistemological practice otherwise depicted as a body-defying phantasmagoria; see his *Sound Unseen*, 99–101.

37. Magee, *Philosophy of Schopenhauer*, 258–59.

38. Ghosh, *Great Derangement*, 120.

39. While the Ideas correspond to the (putatively) real species and categories of natural phenomena, concepts denote human attempts to re-present what exists (or might exist) in the virtual realm of thought. Schopenhauer explains the difference as follows: "The *concept* is abstract, discursive . . . attainable and intelligible only to him who has the faculty of reason, communicable by words without further assistance, entirely exhausted by its definition. The *Idea*, on the other hand, definable perhaps as the adequate representative of the concept, is absolutely perceptive, and, although representing an infinite number of individual things, is yet thoroughly definite" (*WWR*, 1:234).

40. Schopenhauer continues, "Now since the faculty of reason is given to all, but power of judgment to few, the consequence is that man is exposed to delusion, since he is abandoned to every conceivable chimera into which he is talked by anyone, and which, acting as motive to his willing, can induce him to commit perversities and follies of all kinds, and to indulge in the most unheard-of extravagances, even in actions most contrary to his animal nature" (*WWR*, 2:69).

41. Compare Gernot Böhme's call for an ecologically oriented "transformation of aesthetics . . . accomplished by and through overcoming the modern self-conception of humanity itself" ("Aesthetics of Nature," 131). This essay is a translation of "Naturästhetik—Eine Perspective?" the first chapter in Böhme, *Für eine ökologische Naturästhetik*.

42. For the original German, see Schopenhauer, *Die Welt als Wille und Vorstellung*, 2:470.

43. On Kant, see chapter 2. On the coproductive "union of art and nature" in the English garden, see Böhme, "Aesthetics of Nature," 130.

44. Of course, humans do take part in the production of natural beauty when they engage in plant breeding and other such horticultural pursuits; the results might be described, in Kantian terms, as "dependent" rather than "free" beauty.

45. Ghosh, *Great Derangement*, 120. Ghosh is discussing literature; one can imagine a different reading of musical modernism, which has often attempted to stamp out subjectivity and mimic certain aspects of natural processes.

46. Böhme, "Aesthetics of Nature," 127, 129.

47. In his book *The End of Nature*, Bill McKibben writes, "By changing the weather, we make

every spot on earth man-made and artificial. We have deprived nature of its independence, and that is fatal to its meaning. Nature's independence *is* its meaning; without it there is nothing but us" (58).

Chapter Four

1. Newcomb, "Schumann and the Marketplace," 267.
2. John Daverio, *Robert Schumann: Herald of a "New Poetic Age*," 177.
3. Robert Schumann to Henriette Voigt, August 11, 1839, in R. Schumann, *Robert Schumann's Briefe*, 144. In the early stages of composing the piece, Schumann did not feel that he had gotten off to a very good start; at the end of the first draft of the opening section, he wrote, "Set down in the absence of better ideas." See R. Schumann, *Arabeske Op. 18, Blumenstück Op. 19*, ii, 16. Translations from German and French sources in this article are my own unless otherwise noted.
4. Robert Schumann to Ernst Becker, August 15, 1839, cited in Daverio, *Robert Schumann*, 177. The original reads "schwächlich und für Damen"; see R. Schumann, *Robert Schumanns Briefe*, 2nd ed., 169. On piano music written "for ladies," see Solie, "Girling at the Parlor Piano," chap. 3 of her *Music in Other Words*.
5. Daverio, *Robert Schumann*, 177.
6. Reiman writes that *Blumenstück* is "connected by a tissue of intratextual references that mark the work as typical 1830s Schumann, despite its comparative harmonic and textural simplicity" (*Schumann's Piano Cycles*, 172).
7. Jean Paul, *Horn of Oberon*, 70.
8. Robert Schumann to Clara Wieck, January 24, 1839, in R. Schumann, *Jugendbriefe von Robert Schumann*, 297–98.
9. Daverio, *Robert Schumann*, 176–77.
10. See Seaton, *Language of Flowers*, 66–68.
11. Reiman, *Schumann's Piano Cycles*, 159; Jean Paul, *Horn of Oberon*, 186–87, 182–84. In contrast to the humble social settings of "Dutch" novels, "German" and "Italian" novels feature bourgeois and noble characters, respectively.
12. Jean Paul, at any rate, classed *Siebenkäs* among his "German" novels.
13. Parker and Pollock, *Old Mistresses*, 51.
14. Chadwick, *Women, Art, and Society*, 117–18.
15. Ibid., 132.
16. Ibid., 118.
17. Alpers, "Art History and Its Exclusions."
18. Ibid., 187.
19. Chadwick, *Women, Art, and Society*, 138.
20. Ibid., 146–48, 161–74.
21. Jean Paul, *Horn of Oberon*, 135.
22. Hegel, *Aesthetics*, 1:400.
23. Parker and Pollock, *Old Mistresses*, 13.
24. R. Schumann, *Jugendbriefe*, 282–83; translation adapted from R. Schumann, *Early Letters of Robert Schumann*, 270–71. Lawrence Kramer discusses another instance of floral imagery in Schumann's reviews; namely, Schumann's comparison of Franz Schubert's waltzes with roses whose thorns draw blood—an image with curiously mixed gendered connotations. See Kramer, *Franz Schubert*, 97–98.
25. Daverio, *Robert Schumann*, 177.

26. The poems come from Heine's *Lyrisches Intermezzo*, first published in 1823 as part of the *Tragödien, nebst einem lyrischen Intermezzo* and later in the *Buch der Lieder* (1827).

27. R. Schumann, "Vierter und fünfter Quartett-Morgen," 51.

28. I provide a more detailed discussion of the workings of metaphor in this passage in my book *Metaphors of Depth*, chap. 3, "Robert Schumann and Poetic Depth."

29. Jean Paul, *Horn of Oberon*, 130.

30. Seaton, *Language of Flowers*, 51.

31. Novalis [Friedrich von Hardenberg], *Henry von Ofterdingen*, , 15.

32. Kittler, *Discourse Networks 1800/1900*, 73.

33. Ibid., 26.

34. Novalis, *Henry von Ofterdingen*, 17.

35. Hoffmann, *Selected Writings*, 1:54.

36. Ibid.

37. Ibid., 58.

38. Kittler, *Discourse Networks*, 78; Hoffmann, "Golden Pot."

39. Motraye's name appears in different spellings at various points in the historical record; I have adopted the version given on the title page of Aubry De La Motraye, *Travels through Europe, Asia, and into Part of Africa*. See also Lady Mary Wortley Montagu, *Turkish Embassy Letters*, 62.

40. Montagu, *Turkish Embassy Letters*, 122; see also Motraye, *Travels*, 254.

41. Montagu, *Turkish Embassy Letters*, 120; Motraye, *Travels*, 404.

42. Goethe's poem "Code" (*Geheimschrift*), from the *West-östlicher Divan*, praises the "secret double script" as follows: "My mistress sweet this treasure, / Her cipher, did impart / In which I find such pleasure / As she devised the art . . . What pious custom tended / I thus reveal to you, / If you have comprehended / Keep quiet and use it too." Goethe, *Poems of the West and East*, 333–34. Heine's description of Goethe's collection in "The Romantic School" is revealing: "it is a *salaam* sent by the Occident to the Orient, and there are strange flowers in it" (*Romantic School*, 43).

43. Novalis, *Henry von Ofterdingen*, 163 (translation slightly altered).

44. The essay appears in the multilingual journal *Fundgruben des Orients*: see Hammer, "Sur le langage des fleurs"; see also Seaton, *Language of Flowers*, 61–63. After studying oriental languages in Vienna, Hammer (1774–1856) enjoyed positions as a diplomat, member of the Aulic Council, and president of the Austrian Academy of Sciences. Today he is known for being more prolific than accurate in his translations and interpretations of Arabic, Persian, and Turkish texts.

45. Hammer, "Sur le langage des fleurs," 34.

46. Anonymous, *Allgemeine Blumensprache nach der neuesten Deutung*, 28, 30.

47. The cycle, composed in 1840, sets several texts replete with botanical imagery, including Julius Mosen's "Der Nussbaum," Heine's "Die Lotosblume" and "Du bist wie eine Blume," and Friedrich Rückert's "Aus den östliche Rosen." Schumann's inclusion of several poems from Goethe's *West-östlicher Divan* may also be taken as an oblique reference to the Eastern origins of *Blumensprache*.

48. The full title of Ewaldt's book is *Neueste Blumensprache: Gedrängter Auszug der neuesten Schriften darüber; Eine Frühlingsgabe für das schöne Geschlecht* (*The Most Recent Language of Flowers: Condensed Extracts from the Latest Writings; A Springtime Gift for the Fair Sex*).

49. Anonymous, *Neue vollständige Blumensprache*, 3.

50. Charlotte de Latour was the pen name of an author of uncertain identity; she is gener-

ally supposed to have been Louise Cortambert, wife of a French physician and mother of the geographer Eugène Cortambert. See Seaton, *Language of Flowers*, 70–72.

51. C. Latour, *Die Blumensprache, oder Symbolik des Pflanzenreichs*, v–vi.

52. Ibid., xvii.

53. Symanski, *Selam, oder die Sprache der Blumen*, 2nd ed., 57. Symanski's book appears to have been first published in 1820.

54. Miller's poem ends, "O Mädchen, jung und schön von Gesicht, / Vertraue dich dem Flatterer nicht; / Er pflückt der Unschuld Blüte dir ab, / Und bringt dich bald hohnlächelnd ins Grab." See Symanski, *Selam*, 533. Platner's authorship of "Blumensprache," a poem that appeared in an 1805 *Taschenbuch* accompanied solely by the letters "Pl.," is uncertain. See Cline, *Schubert and His World*, 154.

55. Kittler, *Discourse Networks*, 124.

56. Man, *The Rhetoric of Romanticism*, 6.

57. Ibid., 3.

58. As if to correct the form of Schumann's piece, Vladimir Horowitz inserts section I between the last iteration of section II and the four-measure coda in his 1975 Carnegie Hall recording of *Blumenstück* (BMG 82876-50749-2).

59. Kaminsky, "Principles of Formal Structure in Schumann's Early Piano Cycles," 208.

60. Schor, *Reading in Detail*, 3–5.

61. Arnfried Edler also notes the appearance of the BACH motive; see his commentary on *Blumenstück* in "Werke für Klavier zu zwei Händen bis 1840."

62. Daverio, *Robert Schumann*, 167. See also Marston, "Schumann's Heroes," 56.

63. Kittler, *Discourse Networks*, 42–43.

64. On Friedrich Schlegel's references to combination and their relevance to Schumann, see my *Metaphors of Depth*, chap. 3; and Daverio, *Nineteenth-Century Music*, chap. 3.

65. See the diminutive bunch of cornflowers Johannes Brahms dried, pressed, and mounted in a book gifted to Robert during his final illness (C. Schumann, *Blumenbuch für Robert*, 23). Brahms presented the book to Clara on the day of her last performance in Hamburg; she added numerous additional specimens.

66. Jean Paul, *Flower, Fruit and Thorn Pieces*, 340.

Chapter Five

1. Horkheimer and Adorno, *Dialectic of Enlightenment*, 33–34.

2. Derrida, "And Say the Animal Responded?"

3. Ibid., 130.

4. Ibid., 128.

5. Ibid., 127.

6. Ibid., 128. Affect theory has its roots in the physiologically oriented work of psychologist Silvan Tomkins (see the four volumes of his *Affect, Imagery, Consciousness*). Much of what is currently pursued under the rubric of affect theory in the humanities, however, makes little reference to empirical studies in physiology or psychology; it more often takes its bearings from certain strains of cultural studies (say, the work of Raymond Williams, e.g., *Keywords*) or philosophy (especially Gilles Deleuze and Félix Guattari's *A Thousand Plateaus*). See, for example, the diverse set of approaches collected in Gregg and Seigworth, *Affect Theory Reader*. For a critical perspective on affect theory, see Leys, "Turn to Affect"; and Frank et al., "Critical Response."

7. Waal, *Our Inner Ape*.

8. Haraway, *When Species Meet*, 22.

9. Derrida, "And Say the Animal Responded?" 128.

10. Hanslick, *On the Musically Beautiful*, 58 (in this chapter hereafter cited in text as *OMB*); Theodor Adorno, "On the Fetish Character in Music," 292.

11. Thomas Nagel posits that humans are creatures "who can control their actions in response to reasons"; the implication is that we are the only species who can do so. See his *Mind and Cosmos*, 117.

12. Derrida, "And Say the Animal Responded?" 139 (emphasis in original).

13. Horkheimer and Adorno, *Dialectic of Enlightenment*, 33.

14. Baker and Christensen, *Aesthetics and the Art of Musical Composition in the German Enlightenment*, 82. For a recent perspective on music's close ties to movement, see Phillips-Silver, "On the Meaning of Movement."

15. In *The Birth of Tragedy* (1872), Nietzsche wrote, "The music of Apollo was Doric architectonics in tones. . . . The very element which forms the essence of Dionysian music (and hence of music in general) is carefully excluded as un-Apollinian—namely, the emotional power of the tone" (*Birth of Tragedy*, 40).

16. Adorno, "On the Fetish Character in Music," 270; see also Adorno, "Über den Fetischcharakter in der Musik," 14.

17. Sulzer quoted in Baker and Christensen, *Aesthetics and the Art of Musical Composition*, 81.

18. Ibid.

19. Ibid., 82.

20. Riley, *Musical Listening in the German Enlightenment*, chap. 3, "Sulzer and the Aesthetic Force of Music." See also Hoyt, "On the Primitives of Music Theory."

21. Baker and Christensen, *Aesthetics and the Art of Musical Composition*, 91.

22. Ibid., 93.

23. Ibid., 85.

24. Ibid. Agesilaus II was a Spartan king whose exploits were described by Plutarch; see ibid., 85n1.

25. Ibid. Derrida places much of the blame for this view on René Descartes (Derrida, "And Say the Animal Responded?" 121, 143). A century later, the influential Swiss physiologist Albrecht von Haller limited "sensibility" in animals ("in whom the existence of a soul is not so clear") to their capacity to experience pain, ascribing the rest of their seeming responsiveness to mere "irritability" (Vila, *Enlightenment and Pathology*, 21).

26. J-J Rousseau and Herder, *On the Origin of Language*, 87.

27. Ibid.

28. Ibid., 91.

29. See Herder, *Selected Writings on Aesthetics*, 368.

30. Baker and Christensen, *Aesthetics and the Art of Musical Composition*, 90.

31. Ibid., 85.

32. Nina Noeske has explored the gender bias evident in Hanslick's subordination of the body to the mind; see her "Body and Soul, Content and Form."

33. For the original German passage, see Dietmar Strauß, *Eduard Hanslick*, 136.

34. See Karnes, *Music, Criticism, and the Challenge of History*, part 1.

35. Goethe, *Scientific Studies*, 300.

36. Ibid., 301. Compare Nils Wallin's discussion of the polarity between "relaxation" and "strain" in the brain's homeostatic states, from which he extrapolates, in the case of music, a "ba-

sic isomorphy between recurrent, specific physiological processes and recurrent, specific sound gestures" (*Biomusicology*, 483–84).

37. Adorno, *Aesthetic Theory*, 78 (quoted in chapter 1).

38. In a recent position piece, Marcello Sorce Keller suggests that by placing musicology and ethnomusicology under the rubric of zoomusicology (the study of the "organized sounds" of animals), we may come to know ourselves, and our distant animal kin, better. See his "Zoomusicology and Ethnomusicology."

39. Judith Becker writes, "All the studies which locate the reception of music in special areas of the frontal lobe, that is, melody in the right, harmony in the right, rhythm, singing, and syntactic sequences in the left ignore the massive reentrant linkages which trigger neuronal activity in many specialized and distant brain areas" (*Deep Listeners*, 114). On the broadly physiological aspects of music making, see, for example, Peretz, "The Nature of Music from a Biological Perspective," in Peretz et al., "The Nature of Music," 1–32; and Zatorre and Peretz, *Biological Foundations of Music*.

40. Phillips-Silver and Trainor, "Hearing What the Body Feels," 543.

41. Phillips-Silver, "On the Meaning of Movement," 300, 305–6. Wallin adds the visual system to this list (*Biomusicology*, 22).

42. Toiviainen and Keller, "Spatiotemporal Music Cognition," 1.

43. Sedlmeier, Weigelt, and Walther, "Music Is in the Muscle," 297.

44. Ibid., 303.

45. Herder, *Kalligone* (1800), 34.

46. Becker, *Deep Listeners*, 8.

47. Ibid., 7.

48. Ibid., 49. See also Ellis and Thayer, "Music and Autonomic Nervous System (Dys)Function"; they remark that "humans interact with music, both consciously and unconsciously, at behavioral, emotional, and physiological levels" (323).

49. Ibid., 52.

50. Schäfer and Sedlmeier, "Does the Body Move the Soul?"

51. DeNora, *Music Asylums*, 105.

52. Cox, "Embodying Music."

53. Nietzsche, *Will to Power*, 270, 422.

54. Ibid., 422, 434.

55. Ibid., 439n145, 422.

56. Ibid., 285–86.

57. Ibid., 422.

58. Deleuze and Guattari, *A Thousand Plateaus*, 274.

59. Ibid., 275.

60. Ibid., 301.

61. Grosz, *Incorporeal*, 236, 238. Here Grosz is paraphrasing the philosophy of Raymond Ruyer, for whom "mnemic themes" signify the patterns of energy manifested in naturally occurring organic and inorganic entities.

62. Deleuze and Guattari, *A Thousand Plateaus*, 275, 304.

63. Ibid., 308.

64. Grosz, *Chaos, Territory, Art*, 35 (here Grosz follows Deleuze and Guattari: see ibid., 1n9); Herder, "A Monument to Baumgarten," in *Selected Writings on Aesthetics*, 44.

65. Grosz, *Chaos, Territory, Art*, 11; see also Prum, *Evolution of Beauty*.

66. On this topic, see Tomlinson, *A Million Years of Music*; David Huron, "Is Music an Evolutionary Adaptation?"; and Wallin, Merker, and Brown, *Origins of Music*.

67. Herder, *Kalligone*, 34. Incidentally, Herder's latter claim is false; many species, for example, respond to the alarm calls of others, while dogs have proven themselves experts at responding to the sounds of humans.

68. Ibid., 36. Herder's phrase, "die *Mitstimme* der Empfindenden," might also be translated as "the *fellow voice* of the one who is feeling." See Herder, *Herders Werke*, 18:601.

69. See Chaverri, Gillam, and Kunz, "Call-and-Response System."

70. Geissmann, "Gibbon Songs and Human Music," 119.

71. Marler, "Origins of Music and Speech," 31.

72. In chapter 6, I will explore another meaning of the symbolic that likely encompasses music.

73. Marler, "Origins of Music and Speech," 39.

74. Ibid., 40. For a more in-depth consideration of birdsong, see Marler and Slabbekoorn, *Nature's Music*.

75. See Waal, *Are We Smart Enough*, 109.

76. Merker, "Vocal Learning Constellation."

77. Ibid., 221.

78. Ackerman, *Genius of Birds*, 144–45.

79. Merker, "Vocal Learning Constellation," 222.

80. Marler, "Origins of Music and Speech," 40.

81. Ackerman, *Genius of Birds*, 154; Hegel, *Philosophy of Nature*, 409. Some birds also appear to dance for the immediate enjoyment of self. See the study of Snowball, the sulphur-crested cockatoo, by Patel et al., "Experimental Evidence for Synchronization to a Musical Beat in a Nonhuman Animal."

82. Hartshorne, *Born to Sing*, 46. On the twentieth-century decline of interest in aesthetic interpretations of birdsong, see Mundy, "Birdsong and the Image of Evolution."

83. Ackerman, *Genius of Birds*, 143.

84. Grosz, *Chaos, Territory, Art*, 39.

85. Tomlinson, *A Million Years of Music*, 117.

86. Ibid., 259.

87. W. Tecumseh Fitch takes such a boldly comparative approach in his "Biology and Evolution of Music."

88. Higgins, *Music between Us*, 18.

89. Ibid., 33.

90. "Men learnt to mimic with their mouths the trilling notes of birds long before they were able to enchant the ear by joining together in *tuneful song*." Lucretius, *On the Nature of the Universe*, 213.

91. Quoted in Head, "Birdsong and the Origins of Music," 3.

92. Ibid., 12. The quotation comes from Hawkins's 1776 publication *A General History of the Science and Practice of Music*. For a contemporary formulation of this view, see Krause, *Great Animal Orchestra*.

93. Busby, *General History of Music*, 4.

94. Krause ponders the latter possibility in his *Great Animal Orchestra*.

95. Head, "Birdsong and the Origins of Music," 23.

96. See Mabey, *Cabaret of Plants*, 19–20.

97. Tomlinson makes only a passing reference to the possibility that musical instruments may have been used to "discover or imitate nonhuman sounds" (*A Million Years of Music*, 269).

Chapter Six

1. Adams, "Resonance of Place," 9.
2. Ibid., 10, 15.
3. Ibid., 16.
4. Ibid., 17. Compare Johann Gottfried Herder's claim in *Kalligone* (1800) that "*everything that sounds in nature is music*" (emphasis in original) because all sounds move human listeners (Herder, *Kalligone*, 35). See also Rothenberg, "Introduction: Does Nature Understand Music?" and Krause, *Great Animal Orchestra*.
5. On the shifting configurations of the art–nature relationship especially in the twentieth century, see Goehr, *Elective Affinities*, chaps. 3 and 4.
6. Adams, "Resonance of Place," 16–17.
7. Cage, *Silence*, 10.
8. Ibid., 9.
9. Piekut, "Chance and Certainty," 140.
10. See B. Latour, *We Have Never Been Modern*.
11. For a brief overview of the semiotics of Charles Sanders Peirce, see Atkin, "Peirce's Theory of Signs."
12. Adams, "Resonance of Place," 17.
13. Cage, *Silence*, 10.
14. Adams, "Resonance of Place," 16–17.
15. See Abbate, "Music—Drastic or Gnostic?"
16. Adams, "Resonance of Place," 17.
17. Kohn, *How Forests Think*, 75.
18. Goehr, *Elective Affinities*, 94.
19. Deleuze and Guattari, *A Thousand Plateaus*, 239.
20. Kohn, *How Forests Think*, 39.
21. Adams, "The Place Where You Go to Listen," 181.
22. See Bregman, *Auditory Scene Analysis*.
23. Kohn, *How Forests Think*, 32.
24. Marler, "Origins of Music and Speech," 32. See also Thom, "From the Icon to the Symbol."
25. Waal, *Are We Smart Enough to Know How Smart Animals Are?* 107–8. Terrence Deacon argues against the symbolic nature of such calls in his *Symbolic Species*.
26. On the history of acousmatic sound, see Kane, *Sound Unseen*.
27. See Taruskin, "No Ear for Music," 275.
28. Rothenberg, "Introduction: Does Nature Understand Music?" 6.
29. See Rothenberg, *Survival of the Beautiful*; and Martinelli, "Symptomatology of a Semiotic Research," 264.
30. Shannon, "Mathematical Theory of Communication."
31. Compare Gernot Böhme's attempt to revitalize aesthetics as a discourse opposed to "disciplines which treat the human senses merely as information-processing systems" ("Aesthetics of Nature," 131); see also Böhme, *Für eine ökologische Naturästhetik*.
32. Cage, *Silence*, 10.

33. Cage quoted in Piekut, "Chance and Certainty," 142.

34. This is not to say that we have anything like transparent access to the world; what we take to be present is necessarily shaped by our perceptual and cognitive faculties as well as by technological apparatus.

35. See Sontag, *Against Interpretation*; and Abbate, "Music—Drastic or Gnostic?"

36. See Meillassoux, *After Finitude*; and Harman, *Tool-Being*.

37. Bateson, *Steps to an Ecology of Mind*, 434.

38. Rothenberg, "Introduction: Does Nature Understand Music?" 5–6. Composer David Dunn charts an alternative creative path when he writes, "As distinct from John Cage who wanted to decontextualize sounds so as to 'allow them to be themselves,' I have focused on the recontextualization of the sounds of nature as evidence of purposeful-minded systems" ("Nature, Sound Art, and the Sacred," 100).

39. Nothing, that is, other than the design of the auditory system itself, which, as Dale Purves remarks, does not actually "reveal the physical world." He continues, "What we hear in response to a sound signal—including musical tones—are percepts determined by the history of biological success rather than the physical characteristics of signal sources or local pressure changes at the ear" (*Music as Biology*, 12). On phenomenological notions of reduced listening, see Kane, *Sound Unseen*.

40. Kohn, *How Forests Think*, 42.

41. D. Dunn, "Nature, Sound Art, and the Sacred," 98. For a more sustained treatment of this idea, see Pijanowski et al., "Soundscape Ecology."

42. Adams, "Resonance of Place," 17.

43. Tomlinson, "Parahuman Wagnerism," 191, 194.

44. Ibid., 194–95.

45. Ibid., 197. For a further exploration of biosemiotics in relation to affect theory, object-oriented ontology, and speculative realism, see Tomlinson, "Sign, Affect, and Musicking."

46. Ibid., 197; Tomlinson applies these adjectives to indices.

47. Christopher Norris raises similar objections in his response to Tomlinson's article, "Small Change When We Are to Bodies Gone?"

48. N. Cumming, *Sonic Self*, 91.

49. Ibid., 93.

50. Tomlinson, *A Million Years of Music*, 268.

51. Tomlinson, "Parahuman Wagnerism," 196, 195.

52. Ibid., 193.

53. Ibid., 193.

54. Tomlinson, *A Million Years of Music*, 205.

55. Ibid., 258.

56. Ibid., 193.

57. See Grosz, *Chaos, Territory, Art*.

58. Tomlinson, "Parahuman Wagnerism," 196.

59. Ibid.

60. Kohn, *How Forests Think*, 52.

61. Deacon, *Incomplete Nature*, 374, 372. Tomlinson, by contrast, argues that information "must not be confused with reference and meaning" (*A Million Years of Music*, 140).

62. Uexküll, *A Foray into the Worlds of Animals and Humans, with A Theory of Meaning*, 151. The quote comes from *A Theory of Meaning* (originally published as *Bedeutungslehre*, 1940).

63. According to Albert Atkin, "Peirce was aware that any single sign may display some combination of iconic, indexical, and symbolic characteristics" ("Peirce's Theory of Signs").

64. N. Cumming, *Sonic Self*, 86.

65. Moreover, what constitutes a musical sign in the first place is much less determinate, much less bounded, and much less discrete than in language; for example, each tone of a chord, each chord of a harmonic progression, or an entire passage of music containing those elements (as in the presentation of a topic) might all be considered signs. See Dougherty, "Musical Semeiotic."

66. Kohn, *How Forests Think*, 32. Compare David Lidov, who writes that "the apprehension of relations or rules or regularities, such as a tonality, [gives] rise to *symbols*" in a Peircean sense (*Is Language a Music?* 125).

67. Purves, for instance, argues that major and minor thirds correspond to prominent spectra of excited and subdued human speech (*Music as Biology*, chap. 7).

68. Adams, "Resonance of Place," 16.

69. Kohn, *How Forests Think*, 185.

70. Ibid., 56.

71. The assumption that a continuous pitch spectrum characterizes natural sounds as opposed to human music has always been rather strange, considering that many birdsongs depend on distinct, if not mathematically precise, divisions of their melodic tessitura. It is also an assumption that has been made by composers at opposite ends of the philosophical spectrum. In a 1987 interview, Pierre Schaeffer, the originator of musique concrète, distinguishes between sound as the "vocabulary of nature" and music as the province of "values" that arise only within the context of a system. Glancing back over his compositional career, Schaeffer remarked that "it took me forty years to conclude that nothing is possible outside DoReMi. In other words, I wasted my life" (quoted in Hodgkinson, "Interview with Pierre Schaeffer," 37–38 and 35).

72. It is not clear from Winderen's website how many of the sounds were originally in the ultrasound range.

73. See Thomas Nagel's classic essay "What Is it Like to Be a Bat?"

74. Adams, "Resonance of Place," 17.

75. D. Dunn, "Nature, Sound Art, and the Sacred," 98.

76. Compare Dunn's 2008 composition *Listening to What I Cannot Hear*, an assemblage of ultrasonic sounds made by human artifacts and living things ranging from a stopwatch and wind chimes to bats and trees.

77. See Westerkamp, "Linking Soundscape Composition and Acoustic Ecology"; and Drever, "Soundscape Composition." Drever cites Michel Chion, who, commenting on composers associated with the Groupe de Recherches Musicales, criticized the "fetishisms which have to do principally with focusing attention on the sources of the sounds and the means whereby they are produced, whereas the sounds themselves are what really count" ("Soundscape Composition," 21).

78. Westerkamp, "Linking Soundscape Composition," 52; Truax, "Soundscape, Acoustic Communication and Environmental Sound Composition," 52.

79. Westerkamp, "Speaking from Inside the Soundscape," 150.

80. Ibid., 148.

81. Ibid., 149.

82. Ibid.

83. Ibid., 151.

84. Drever, "Soundscape Composition," 22.//
85. Westerkamp, "Linking Soundscape Composition," 54.
86. Westerkamp, "Speaking from Inside the Soundscape," 148.
87. Ibid., 151.
88. For a brief consideration of place and space in relation to music, see my essay "Musical Ecologies of Place and Placelessness."
89. For a more detailed consideration of the piece, see Feisst, "Music as Place, Place as Music"; see also Morris, "Ecotopian Sounds."
90. Adams, "Resonance of Place," 16.
91. Adams, "Winter Music," 75.
92. Grosz, *Chaos, Territory, Art*, 58. Incidentally, Deleuze and Guattari reject the ideology of sounds themselves: "We are not at all arguing for an aesthetics of qualities, as if the pure quality (color, sound, etc.) held the secret of a becoming without measure." Instead, they claim that "it is through a system of melodic and harmonic coordinates by means of which music reterritorializes upon itself, *qua* music" (*A Thousand Plateaus*, 306, 303).
93. Kohn, *How Forests Think*, 51.
94. Ibid., 177.
95. Tomlinson, "Parahuman Wagnerism," 197.
96. I thank Eric Drott for this suggestion.
97. Harald Krebs discusses this moment in *Fantasy Pieces*, 94.
98. Tunbridge, *Schumann's Late Style*, 194.
99. Kittler, *Discourse Networks*, 42–43.
100. Tunbridge, *Schumann's Late Style*, 191.
101. Kant, *Critique of Judgement*, 142 (translation altered). See also Kant, *Kritik der Urteilskraft*, 300.
102. Cited in Drever, "Soundscape Composition," 22.

Bibliography

Abbate, Carolyn. "Music—Drastic or Gnostic?" *Critical Inquiry* 30, no. 3 (2004): 505–36.
———. *Unsung Voices: Opera and Musical Narrative in the Nineteenth Century.* Princeton, NJ: Princeton University Press, 1991.
Abrams, M. H. *The Mirror and the Lamp: Romantic Theory and the Critical Tradition.* Oxford: Oxford University Press, 1953.
Ackerman, Jennifer. *The Genius of Birds.* New York: Penguin Press, 2016.
Adams, John Luther. "The Place Where You Go to Listen." In Rothenberg and Ulvaeus, *Book of Music and Nature*, 181–82.
———. "Resonance of Place." *North American Review* 279, no. 1 (1994): 8–18.
———. "Winter Music: A Composer's Journal (1998–99)." In *Winter Music: Composing the North*, by John Luther Adams, 57–77. Middletown, CT: Wesleyan University Press, 2004.
Adorno, Theodor W. *Aesthetic Theory.* Translated by Robert Hullot-Kentor. Minneapolis: University of Minnesota Press, 1997.
———. "On the Fetish Character in Music and the Regression of Listening." In *The Essential Frankfurt School Reader*, edited by Andrew Arato and Eike Gebhardt, translated by Maurice Goldbloom, 270–99. New York: Continuum, 1982.
———. "Über den Fetischcharakter in der Musik und die Regression des Hörens." In *Gesammelte Schriften*, vol. 14, edited by Rolf Tiedemann, 14–50. Frankfurt: Suhrkamp, 1997.
———. "Vers une musique informelle." In *Quasi una Fantasia: Essays on Modern Music*, translated by Rodney Livingstone, 269–322. London: Verso, 1998.
Allen, Aaron S. "Ecomusicology from Poetic to Practical." In Zapf, *Handbook of Ecocriticism and Cultural Ecology*, 644–63.
Allen, Aaron S., and Kevin Dawe, eds. *Current Directions in Ecomusicology: Music, Culture, Nature.* New York: Routledge, 2016.
Alpers, Svetlana. "Art History and Its Exclusions: The Example of Dutch Art." In *Feminism and Art History*, edited by Norma Broude and Mary D. Garrard, 183–99. New York: Harper and Row, 1982.
Anonymous. *Allgemeine Blumensprache nach der neuesten Deutung.* Speyer, Ger.: F. C. Neidhard, 1837.

Anonymous. *Neue vollständige Blumensprache. Der Liebe und Freundschaft gewidmet*. 8th ed. Quedlinburg, Ger.: Gottfried Basse, 1850.

Atkin, Albert. "Peirce's Theory of Signs." In *Stanford Encyclopedia of Philosophy*, edited by Edward N. Zalta. Stanford University, 1997–; online ed., summer 2013. Article published October 13, 2006; last modified November 15, 2010. http://plato.stanford.edu/archives/sum2013/entries/peirce-semiotics/.

Atwell, John E. *Schopenhauer on the Character of the World: The Metaphysics of Will*. Berkeley: University of California Press, 1995.

Atzert, Stephan. "Musik und Freiheit vom Willen? Zum reinen Subjekt des Erkennens und der Empfindung beim Musikhören." In *Musik als Wille und Welt: Schopenhauers Philosophie der Musik*, ed. Matthias Koßler, 51–60. Würzburg: Königshausen und Neumann, 2011.

Aucouturier, Jean-Julien. "The Hypothesis of Self-Organization for Musical Tuning Systems." *Leonardo Music Journal* 18 (2008): 63–69.

Baker, Nancy Kovaleff, and Thomas Christensen, eds. *Aesthetics and the Art of Musical Composition in the German Enlightenment: Selected Writings of Johann Georg Sulzer and Heinrich Christoph Koch*. Cambridge: Cambridge University Press, 1995.

Bateson, Gregory. *Steps to an Ecology of Mind*. New York: Ballantine Books, 1972.

Becker, Judith. *Deep Listeners: Music, Emotion, and Trancing*. Bloomington: Indiana University Press, 2004.

Beiser, Frederick. *Hegel*. New York: Routledge, 2005.

Bennett, Jane. *Vibrant Matter: A Political Ecology of Things*. Durham, NC: Duke University Press, 2010.

Bloechl, Olivia, Melanie Lowe, and Jeffrey Kallberg, eds. Introduction to *Rethinking Difference in Music Scholarship*. Cambridge: Cambridge University Press, 2015.

Böhme, Gernot. "Aesthetics of Nature—A Philosophical Perspective." In Zapf, *Handbook of Ecocriticism and Cultural Ecology*, 123–34.

———. *Für eine ökologische Naturästhetik*. Frankfurt: Suhrkamp, 1989.

Bonds, Mark Evan. *Absolute Music: The History of an Idea*. New York: Oxford University Press, 2014.

———. "Aesthetic Amputations: Absolute Music and the Deleted Endings of Hanslick's Vom Musikalisch-Schönen." *19th-Century Music* 36, no. 1 (2012): 3–23.

Bregman, Albert S. *Auditory Scene Analysis: The Perceptual Organization of Sound*. Cambridge, MA: MIT Press, 1999.

Broyles, Michael. "Organic Form and the Binary Repeat." *Musical Quarterly* 66, no. 3 (1980): 339–60.

Buch, Esteban. "Adorno's Schubert: From the Critique of the Garden Gnome to the Defense of Atonalism." *19th-Century Music* 29, no. 1 (2005): 25–30.

Budd, Malcolm. *The Aesthetic Appreciation of Nature: Essays on the Aesthetics of Nature*. Oxford: Clarendon Press, 2002.

Burnham, Scott. *Beethoven Hero*. Princeton, NJ: Princeton University Press, 1995.

———. "How Music Matters: Poetic Content Revisited." In Cook and Everist, *Rethinking Music*, 193–216.

———. "The Second Nature of Sonata Form." In Clark and Rehding, *Music Theory and Natural Order*, 111–41.

Busby, Thomas. *A General History of Music*. New York: Da Capo Press, 1968.

Cage, John. "Experimental Music." In Cage, *Silence*, 7–12.

———. *Silence*. Middletown, CT: Wesleyan University Press, 1961.
Camazine, Scott, Jean-Louis Deneubourg, Nigel R. Franks, James Sneyd, Guy Theraulaz, and Eric Bonabeau. *Self-Organization in Biological Systems*. Princeton, NJ: Princeton University Press, 2001.
Carlson, Allen, and Arnold Berleant. "Introduction: The Aesthetics of Nature." In Carlson and Berleant, *Aesthetics of Natural Environments*, 11–42.
———. *The Aesthetics of Natural Environments*. Peterborough, ON: Broadview Press, 2004.
Cartwright, David E. *Schopenhauer: A Biography*. New York: Cambridge University Press, 2010.
Cecchetto, David. *Humanesis: Sound and Technological Posthumanism*. Minneapolis: University of Minnesota Press, 2013.
Chadwick, Whitney. *Women, Art, and Society*. 3rd ed. London: Thames and Hudson, 2002.
Chakrabarty, Dipesh. "The Climate of History: Four Theses." *Critical Inquiry* 35, no. 2 (2009): 197–222.
Chaverri, Gloriana, Erin H. Gillam, and Thomas H. Kunz. "A Call-and-Response System Facilitates Group Cohesion among Disc-Winged Bats." *Behavioral Ecology* 24, no. 2 (2013): 481–87.
Cherlin, Michael. "Hauptmann and Schenker: Two Adaptations of Hegelian Dialectics." *Theory and Practice* 13 (1988): 115–31.
Chua, Daniel K. L. *Absolute Music and the Construction of Identity*. Cambridge: Cambridge University Press, 1999.
Clark, Suzannah, and Alexander Rehding. *Music Theory and Natural Order from the Renaissance to the Early Twentieth Century*. Cambridge: Cambridge University Press, 2001.
Cline, Peter. *Schubert and His World: A Biographical Dictionary*. Oxford: Clarendon Press, 1997.
Cook, Nicholas. "Theorizing Musical Meaning." *Music Theory Spectrum* 23, no. 2 (2001): 170–95.
Cook, Nicholas, and Mark Everist. *Rethinking Music*. Oxford: Oxford University Press, 1999.
Coole, Diana, and Samantha Frost, eds. *New Materialisms: Ontology, Agency, Politics*. Durham, NC: Duke University Press, 2010.
Cox, Arnie. "Embodying Music: Principles of the Mimetic Hypothesis." *Music Theory Online* 17, no. 2 (2011): 1–24.
Cumming, Julie E. *The Motet in the Age of Du Fay*. Cambridge: Cambridge University Press, 1999.
Cumming, Naomi. *The Sonic Self: Musical Subjectivity and Signification*. Bloomington: Indiana University Press, 2000.
Currie, James. "Music after All." *Journal of the American Musicological Society* 62, no. 1 (2009): 145–203.
———. "There's No Place." *Ecomusicology Review* 21 (2017), https://ethnomusicologyreview.ucla.edu/content/there%E2%80%99s-no-place.
Cusick, Suzanne. "Feminist Theory, Music Theory, and the Mind/Body Problem." *Perspectives of New Music* 32, no. 1 (1994): 8–27.
———. "Musicology, Gender, and Feminism." In Cook and Everist, *Rethinking Music*, 471–98.
———. "'You Are in a Place That Is Out of the World . . .': Music in the Detention Camps of the 'Global War on Terror,'" *Journal of the Society for American Music* 2, no. 1 (2008): 1–26.
Dahlhaus, Carl. *Foundations of Music History*. Translated by J. B. Robinson. Cambridge: Cambridge University Press, 1983.
———. *The Idea of Absolute Music*. Translated by Roger Lustig. Chicago: University of Chicago Press, 1989.

Danielsen, Anne. *Presence and Pleasure: The Funk Grooves of James Brown and Parliament*. Middletown, CT: Wesleyan University Press, 2006.

Daverio, John. *Nineteenth-Century Music and the German Romantic Ideology*. New York: Schirmer Books, 1993.

———. *Robert Schumann: Herald of a "New Poetic Age."* New York: Oxford University Press, 1997.

Deacon, Terrence. *Incomplete Nature: How Mind Emerged from Matter*. New York: W. W. Norton, 2012.

———. *The Symbolic Species: The Co-evolution of Language and the Brain*. New York: W. W. Norton, 1997.

DeLanda, Manuel. *A New Philosophy of Society: Assemblage Theory and Social Complexity*. London: Continuum, 2006.

Deleuze, Gilles, and Félix Guattari. *A Thousand Plateaus: Capitalism and Schizophrenia*. Translated by Brian Massumi. Minneapolis: University of Minnesota Press, 1987.

DeNora, Tia. *Music Asylums: Wellbeing through Music in Everyday Life*. Farnham, UK: Ashgate, 2013.

Derrida, Jacques. "And Say the Animal Responded?" in *Zoontologies: The Question of the Animal*, edited by Cary Wolfe, 121–46. Minneapolis: University of Minnesota Press, 2003.

———. "White Mythology." In *Margins of Philosophy*, translated by Alan Bass, 207–71. Chicago: University of Chicago Press, 1982.

Descola, Philippe. *Beyond Nature and Culture*. Translated by Janet Lloyd. Chicago: University of Chicago Press, 2013.

Dougherty, William P. "Musical Semeiotic: A Peircean Perspective." *Contemporary Music Review* 16, no. 4 (1997): 29–39.

Draheim, Joachim. "Arabeske für Klavier op. 18; Blumenstück für Klavier op. 19." In *Robert Schumann: Interpretationen seiner Werke*, vol. 1, edited by Helmut Loos, 106–10. Laaber, Ger.: Laaber-Verlag, 2005.

Drever, John Levack. "Soundscape Composition: The Convergence of Ethnography and Acousmatic Music." *Organised Sound* 7, no. 1 (2002): 21–27.

Duerksen, Marva. "Schenker's Organicism Revisited." *Intégral* 22 (2008): 1–58.

Dunn, David. "Nature, Sound Art, and the Sacred." In Rothenberg and Ulvaeus, *Book of Music and Nature*, 95–107.

Dunn, Rob. *The Wild Life of Our Bodies: Predators, Parasites, and Partners That Shape Who We Are Today*. New York: HarperCollins, 2011.

Edler, Arnfried. "*Arabeske* op. 18, *Blumenstück* op. 19." In Tadday, *Schumann Handbuch*, 250–51.

———. "Werke für Klavier zu zwei Händen bis 1840." In Tadday, *Schumann Handbuch*, 214–57.

Edwards, James Rhys. "Critical Theory in Ecomusicology." In Allen and Dawe, *Current Directions in Ecomusicology*, 153–64.

Ellis, Robert J., and Julian F. Thayer. "Music and Autonomic Nervous System (Dys)Function." *Music Perception* 27, no. 4 (2010): 317–26.

Ewaldt, Karl Wilhelm. *Neueste Blumensprache: Gedrängter Auszug der neuesten Schriften darüber; Eine Frühlingsgabe für das schöne Geschlecht*. Leipzig: F. A. Serig, 1825.

Feisst, Sabine. "Music as Place, Place as Music: The Sonic Geography of John Luther Adams." In *The Farthest Place: The Music of John Luther Adams*, edited by Bernd Herzogenrath, 23–47. Boston: Northeastern University Press, 2012.

Feisst, Sabine, Denise Von Glahn, Ellen Waterman, and Garth Paine, eds. "Music and Ecology." Special issue, *Contemporary Music Review* 35, no. 3 (2016).

Felski, Rita. *The Limits of Critique*. Chicago: University of Chicago Press, 2015.
Ferrara, Lawrence. "Schopenhauer on Music as the Embodiment of Will." In Jacquette, *Schopenhauer, Philosophy, and the Arts*, 183–99.
Fink, Robert. "Goal-Directed Soul? Analyzing Rhythmic Teleology in African American Popular Music." *Journal of the American Musicological Society* 64, no. 1 (2011): 179–231.
Fitch, W. Tecumseh. "The Biology and Evolution of Music: A Comparative Perspective." *Cognition* 100, no. 1 (2006): 173–215.
Frank, Adam, Elizabeth A. Wilson, Charles Altieri, and Ruth Leys. "Critical Response." *Critical Inquiry* 38, no. 4 (2012): 870–91.
Gagliano, Monica. "Green Symphonies: A Call for Studies on Acoustic Communication in Plants." *Behavioral Ecology* 24, no. 4 (2013): 789–96.
———. "Singing Plants at Damanhur." Video. Accessed via YouTube March 17, 2018. https://www.youtube.com/watch?v=aZaokNmQ4eY.
Gallope, Michael. *Deep Refrains: Music, Philosophy, and the Ineffable*. Chicago: University of Chicago Press, 2017.
Gautier, Ana María Ochoa. "Acoustic Multinaturalism, the Value of Nature, and the Nature of Music in Ecomusicology." *boundary 2* 43, no. 1 (2016): 107–41.
Geissmann, Thomas. "Gibbon Songs and Human Music from an Evolutionary Perspective." In Wallin, Merker, and Brown, *Origins of Music*, 103–23.
Ghosh, Amitav. *The Great Derangement: Climate Change and the Unthinkable*. Chicago: University of Chicago Press, 2016.
Gjerdingen, Robert. *Music in the Galant Style*. Oxford: Oxford University Press, 2007.
Goehr, Lydia. *Elective Affinities: Musical Essays on the History of Aesthetic Theory*. New York: Columbia University Press, 2008.
———. "Schopenhauer and the Musicians." In Jacquette, *Schopenhauer, Philosophy, and the Arts*, 200–228.
Goethe, Johann Wolfgang von. *Poems of the West and East*. Translated by John Whaley. Bern: Peter Lang, 1998.
———. *Scientific Studies*. Edited and translated by Douglas Miller. Vol. 12 of *Collected Works*. Princeton, NJ: Princeton University Press, 1994.
Goodman, Steve. *Sonic Warfare: Sound, Affect, and the Ecology of Fear*. Cambridge, MA: MIT Press, 2010.
Gregg, Melissa, and Gregory J. Seigworth, eds. *The Affect Theory Reader*. Durham, NC: Duke University Press, 2010.
Grier, James. *The Critical Editing of Music: History, Method, and Practice*. Cambridge: Cambridge University Press, 1996.
Grimes, Nicole, Siobhán Donovan, and Wolfgang Marx, eds. *Rethinking Hanslick: Music, Formalism, and Expression*. Rochester, NY: University of Rochester Press, 2013.
Grimley, Daniel. "Organicism, Form, and Structural Decay: Nielsen's Second Violin Sonata." *Music Analysis* 21, no. 2 (2002): 175–205.
Grogan, David, ed. "The Microbiome." Special edition, *Scientific American* 312, no. 3 (March 2015): S1–S16.
Grosz, Elizabeth. *Chaos, Territory, Art: Deleuze and the Framing of the Earth*. New York: Columbia University Press, 2008.
———. *The Incorporeal: Ontology, Ethics, and the Limits of Materialism*. New York: Columbia University Press, 2017.

———. *The Nick of Time: Politics, Evolution, and the Untimely*. Durham, NC: Duke University Press, 2004.

Grusin, Richard, ed. *The Nonhuman Turn*. Minneapolis: University of Minnesota Press, 2015.

Habermas, Jürgen. *The Philosophical Discourse of Modernity: Twelve Lectures*. Translated by Frederick Lawrence. Cambridge, MA: MIT Press, 1987.

Haldane, J. S. *The Philosophy of a Biologist*. 2nd ed. Oxford: Clarendon Press, 1936.

Halperin, David. "What Is Sex For?" *Critical Inquiry* 43, no. 1 (2016): 1–31.

Hammer, Joseph. "Sur le langage des fleurs." In *Fundgruben des Orients*, vol. 1, 32–42. Vienna: Anton Schmid, 1809.

Hannan, Barbara. *The Riddle of the World: A Reconsideration of Schopenhauer's Philosophy*. Oxford: Oxford University Press, 2009.

Hanslick, Eduard. *On the Musically Beautiful*. Edited and translated by Geoffrey Payzant. Indianapolis: Hackett, 1986.

Haraway, Donna J. *Crystals, Fabrics, and Fields: Metaphors of Organicism in Twentieth-Century Developmental Biology*. New Haven, CT: Yale University Press, 1976.

———. *When Species Meet*. Minneapolis: University of Minnesota Press, 2008.

Harman, Graham. *Tool-Being: Heidegger and the Metaphysics of Objects*. Chicago: Open Court, 2002.

Harrington, Anne. *Reenchanted Science: Holism in German Culture from Wilhelm II to Hitler*. Princeton, NJ: Princeton University Press, 1996.

Hartshorne, Charles. *Born to Sing: An Interpretation and World Survey of Bird Song*. Bloomington: Indiana University Press, 1973.

Hatab, Lawrence J. "Nietzsche, Nature, and the Affirmation of Life." In Lemm, *Nietzsche and the Becoming of Life*, 32–48.

Hayles, N. Katherine. "The Cognitive Nonconscious: Enlarging the Mind of the Humanities," *Critical Inquiry* 42, no. 4 (2016): 783–808.

———. *How We Became Posthuman: Virtual Bodies in Cybernetics, Literature, and Informatics*. Chicago: University of Chicago Press, 1999.

Head, Matthew. "Birdsong and the Origins of Music." *Journal of the Royal Musical Association* 122, no. 1 (1997): 1–23.

Hegel, G. W. F. *Aesthetics: Lectures on Fine Art*. 2 vols. Translated by T. M. Knox. Oxford: Clarendon Press, 1998.

———. *Elements of the Philosophy of Right*. Edited by Allen W. Wood. Translated by H. B. Nisbet. Cambridge: Cambridge University Press, 1991.

———. *Philosophy of Nature*. Translated by A. V. Miller. Oxford: Clarendon Press, 2004.

Heine, Heinrich. *The Romantic School and Other Essays*. Edited by Jost Hermand and Robert C. Holub. New York: Continuum, 1985.

Helmreich, Stefan. *Sounding the Limits of Life: Essays in the Anthropology of Biology and Beyond*. Princeton, NJ: Princeton University Press, 2016.

Hepburn, Ronald. "Contemporary Aesthetics and the Neglect of Natural Beauty." In Carlson and Berleant, *Aesthetics of Natural Environments*, 43–62. First published in 1966.

Herder, Johann Gottfried. *Herders Werke*. 24 vols. Edited by Heinrich Düntzer. Berlin: Gustav Hempel, 1879.

———. *Kalligone* (excerpt). In Edward A. Lippman, *Musical Aesthetics: A Historical Reader*, vol. 2, 33–43. Stuyvesant, NY: Pendragon Press, 1988.

———. *Selected Writings on Aesthetics*. Edited and translated by Gregory Moore. Princeton, NJ: Princeton University Press, 2006.

BIBLIOGRAPHY

Hiekel, Jörn Peter, ed. *Ins Offene? Neue Musik und Natur*. Mainz: Schott, 2014.

Higgins, Kathleen Marie. *The Music between Us: Is Music a Universal Language?* Chicago: University of Chicago Press, 2012.

Hodgkinson, Tim. "An Interview with Pierre Schaeffer." In Rothenberg and Ulvaeus, *Book of Music and Nature*, 34–44.

Hoffmann, E. T. A. "E. T. A. Hoffmann: '[Review: Beethoven's Symphony no. 5 in C minor]' (1810)." In *Music Analysis in the Nineteenth Century*, vol. 2, edited by Ian Bent, 141–60. Cambridge: Cambridge University Press, 1994.

———. *E. T. A. Hoffmann's Musical Writings: Kreisleriana, The Poet and the Composer, Music Criticism*. Edited by David Charlton. Translated by Martyn Clarke. Cambridge: Cambridge University Press, 1989.

———. "The Golden Pot." In *Tales*, edited by Victor Lange, 1–79. New York: Continuum, 1982.

———. "Recension" [review of Beethoven, Symphony no. 5]. *Allgemeine musikalische Zeitung* 12, no. 40 (July 4, 1810): 630–42, and no. 41 (July 11, 1810): 652–59.

———. *Selected Writings of E. T. A. Hoffmann*. 2 vols. Edited and translated by Leonard J. Kent and Elizabeth C. Knight. Chicago: University of Chicago Press, 1969.

Holmes, Brooke. "Before Nature?" Foreword to *Ecocriticism, Ecology, and the Cultures of Antiquity*, edited by Christopher Schliephake, ix–xiii. Lanham, MD: Lexington Books, 2017.

Horkheimer, Max, and Theodor W. Adorno. *Dialectic of Enlightenment*. Translated by John Cumming. New York: Continuum, 1994.

Hoyt, Peter A. "On the Primitives of Music Theory: The Savage and Subconscious as Sources of Analytical Authority." In Clark and Rehding, *Music Theory and Natural Order*, 197–212.

Huron, David. "Is Music an Evolutionary Adaptation?" in Zatorre and Peretz, *Biological Foundations of Music*, 43–61.

Jacquette, Dale. *Schopenhauer, Philosophy, and the Arts*. Cambridge: Cambridge University Press, 1996.

Janaway, Christopher. "Will and Nature." In *The Cambridge Companion to Schopenhauer*, edited by Christopher Janaway, 138–70. New York: Cambridge University Press, 1999.

Janz, Tobias. *Zur Genealogie der musikalischen Moderne*. Paderborn, Ger.: Wilhelm Fink, 2014.

Jean Paul. *Flower, Fruit and Thorn Pieces: Or the Married Life, Death, and Wedding of the Advocate of the Poor, Firmian Stanislaus Siebenkäs*. Vol. 1. Translated by Edward Henry Noel. Boston: James Munroe, 1845.

———. *Horn of Oberon: Jean Paul Richter's School for Aesthetics*. Translated by Margaret R. Hale. Detroit: Wayne State University Press, 1973.

Kaminsky, Peter. "Principles of Formal Structure in Schumann's Early Piano Cycles." *Music Theory Spectrum* 11, no. 2 (1989): 207–25.

Kane, Brian. *Sound Unseen: Acousmatic Sound in Theory and Practice*. New York: Oxford University Press, 2014.

Kant, Immanuel. *Critique of Judgement*. Translated by J. H. Bernard. New York: Hafner Press, 1951.

———. *Kritik der Urteilskraft*. Edited by Wilhelm Windelband. In *Gesammelte Schriften*, vol. 5, by Immanuel Kant, edited by Royal Prussian Academy of Sciences. Berlin: Georg Reimer, 1913.

Karnes, Kevin. *Music, Criticism, and the Challenge of History: Shaping Modern Musical Thought in Late Nineteenth-Century Vienna*. New York: Oxford University Press, 2008.

Kawohl, Friedemann. "Organismusmetaphern." In *Musiktheorie*, edited by Helga de la Motte-Haber and Oliver Schwab-Felisch, 156–67. Laaber, Ger.: Laaber-Verlag, 2005.

Keller, Marcello Sorce. "Zoomusicology and Ethnomusicology: A Marriage to Celebrate in Heaven." *Yearbook for Traditional Music* (2012): 166–83.

Kerman, Joseph. *Contemplating Music*. Cambridge, MA: Harvard University Press, 1985.

———. "How We Got Into Analysis and How to Get Out." *Critical Inquiry* 7, no. 2 (1980): 311–22.

Kittler, Friedrich. *Discourse Networks 1800/1900*. Translated by Michael Metteer, with the assistance of Chris Cullens. Stanford, CA: Stanford University Press, 1990.

Kohn, Eduardo. *How Forests Think: Toward an Anthropology Beyond the Human*. Berkeley: University of California Press, 2013.

Kollias, Phivos-Angelos. "The Self-Organising Work of Music." *Organised Sound* 16, no. 2 (2011): 192–99.

Korsyn, Kevin. "Schenker's Organicism Reexamined." *Intégral* 7 (1993): 82–118.

Kramer, Lawrence. *Classical Music and Postmodern Knowledge*. Berkeley: University of California Press, 1995.

———. *Franz Schubert: Sexuality, Subjectivity, Song*. Cambridge: Cambridge University Press, 1998.

Krause, Bernie. *The Great Animal Orchestra: Finding the Origins of Music in the World's Wild Places*. New York: Little, Brown, 2012.

Krebs, Harald. *Fantasy Pieces: Metrical Dissonance in the Music of Robert Schumann*. New York: Oxford University Press, 1999.

Krieger, Murray. *A Reopening of Closure: Organicism Against Itself*. New York: Columbia University Press, 1989.

Langer, Susanne. *Feeling and Form*. New York: Charles Scribner's Sons, 1953.

Large, Edward W. "On Synchronizing Movement to Music." *Human Movement Science* 19, no. 4 (2000): 527–66.

Larson, Steve. "Musical Forces and Melodic Patterns." *Theory and Practice* 22–23 (1997–98): 55–72.

Latour, Bruno. *We Have Never Been Modern*. Translated by Catherine Porter. Cambridge, MA: Harvard University Press, 1993.

Latour, Charlotte de. *Die Blumensprache, oder Symbolik des Pflanzenreichs*. Translated by Karl Müchler. Berlin: Karl August Stuhr, 1820.

LeMenager, Stephanie, Teresa Shewry, and Ken Hiltner, eds. *Environmental Criticism for the Twenty-First Century*. New York: Routledge, 2011.

Lemm, Vanessa. Introduction to Lemm, *Nietzsche and the Becoming of Life*, 1–15.

———. *Nietzsche and the Becoming of Life*. New York: Fordham University Press, 2015.

Lerdahl, Fred, and Ray Jackendoff. *A Generative Theory of Tonal Music*. Cambridge, MA: MIT Press, 1983.

Levy, Janet. "Covert and Casual Values in Recent Writings about Music." *Journal of Musicology* 5, no. 1 (1987): 3–27.

Leys, Ruth. "The Turn to Affect: A Critique." *Critical Inquiry* 37, no. 3 (2011): 434–72.

Lidov, David. *Is Language a Music? Writings on Musical Form and Signification*. Bloomington: Indiana University Press, 2005.

Love, Glen A. *Practical Ecocriticism: Literature, Biology, and the Environment*. Charlottesville: University of Virginia Press, 2003.

Lovejoy, Arthur O. *The Great Chain of Being: A Study of the History of an Idea*. Cambridge, MA: Harvard University Press, 1936.

Lucretius. *On the Nature of the Universe*. Translated by R. E. Latham. London: Penguin Books, 1988.

Luhmann, Niklas. *Art as a Social System*. Translated by Eva M. Knodt. Stanford, CA: Stanford University Press, 2000.

———. "Self-Organization and Autopoiesis." In *Emergence and Embodiment: New Essays on Second-Order Systems Theory*, edited by Bruce Clarke and Mark N. B. Hansen, 143–56. Durham, NC: Duke University Press, 2009.

———. *Social Systems*. Translated by John Bednarz Jr., with the assistance of Dirk Baecker. Stanford, CA: Stanford University Press, 1995.

Mabey, Richard. *The Cabaret of Plants: Forty Thousand Years of Plant Life and the Human Imagination*. New York: W. W. Norton, 2015.

Madden, Charles. *Fractals in Music: Introductory Mathematics for Musical Analysis*. Salt Lake City: High Art Press, 1999.

Magee, Bryan. *The Philosophy of Schopenhauer*. Rev. ed. Oxford: Oxford University Press, 1997.

Malabou, Catherine. "One Life Only: Biological Resistance, Political Resistance." *Critical Inquiry* 42, no. 3 (2016): 429–38.

Man, Paul de. *The Rhetoric of Romanticism*. New York: Columbia University Press, 1984.

Marder, Michael. "The Life of Plants and the Limits of Empathy." *Dialogue* 51 (2012): 259–73.

———. *Plant-Thinking: A Philosophy of Vegetal Life*. New York: Columbia University Press, 2013.

Marder, Michael, and Monica Gagliano. "Michael Marder and Monica Gagliano: How Do Plants Sound?" *Columbia University Press Blog*, June 19, 2013, http://www.cupblog.org/?p=10609.

Marler, Peter. "Origins of Music and Speech: Insights from Animals." In Wallin, Merker, and Brown, *Origins of Music*, 31–48.

Marler, Peter, and Hans Slabbekoorn. *Nature's Music: The Science of Birdsong*. San Diego: Elsevier, 2004.

Marston, Nicholas. "Schumann's Heroes: Schubert, Beethoven, Bach." In *The Cambridge Companion to Schumann*, edited by Beate Perrey, 48–61. Cambridge: Cambridge University Press, 2007.

Martinelli, Dario. "Symptomatology of a Semiotic Research." In *Musical Semiotics Revisited*, edited by Eero Tarasti, 261–71. Imatra, Fin.: International Semiotics Institute, 2003.

Marx, Adolf Bernhard. *Musical Form in the Age of Beethoven: Selected Writings on Theory and Method*. Edited and translated by Scott Burnham. Cambridge: Cambridge University Press, 1997.

Maturana, Humberto R., and Francisco J. Varela. *Autopoiesis and Cognition: The Realization of the Living*. Dordrecht, Neth.: D. Reidel, 1980.

Maus, Fred Everett. "Hanslick's Animism." *Journal of Musicology* 10, no. 3 (1992): 273–92.

———. "Music as Drama." *Music Theory Spectrum* 10 (1988): 56–73.

McClary, Susan. *Conventional Wisdom: The Content of Musical Form*. Berkeley: University of California Press, 2000.

———. *Feminine Endings: Music, Gender, and Sexuality*. Minneapolis: University of Minneapolis Press, 1991.

McCreless, Patrick, Kofi Agawu, Joseph Dubiel, Scott Burnham, Matthew Brown, and Marion A. Guck. "Contemporary Music Theory and the New Musicology." *Journal of Musicology* 15, no. 3 (1997): 291–352.

McKibben, Bill. *The End of Nature*. New York: Random House, 1989.

McLaughlin, Peter. *Kant's Critique of Teleology in Biological Explanation: Antinomy and Teleology*. Studies in the History of Philosophy, vol. 16. Lewiston, NY: Edwin Mellen Press, 1990.

Mead, Andrew. "Cultivating an Air: Natural Imagery and Music Making." *Perspectives of New Music* 52, no. 2 (2014): 91–118.

Meillassoux, Quentin. *After Finitude: An Essay on the Necessity of Contingency.* Translated by Ray Brassier. London: Continuum, 2008.

Merker, Björn. "The Vocal Learning Constellation: Imitation, Ritual Culture, Encephalization." In *Music, Language, and Human Evolution,* edited by Nicholas Bannan, 215–60. Oxford: Oxford University Press, 2012.

Meyer, Leonard. *Music, the Arts, and Ideas: Patterns and Predications in Twentieth-Century Culture.* Chicago: University of Chicago Press, 1967.

———. *Style and Music: Theory, History, and Ideology.* Philadelphia: University of Pennsylvania Press, 1989.

Michaelis, Christian Friedrich. "Ein Versuch, das innere Wesen der Tonkunst zu entwickeln." *Allgemeine musikalische Zeitung* 8, no. 43 (July 23, 1806): 673–83, and no. 44 (July 30, 1806): 691–96.

Miller, Elaine P. *The Vegetative Soul: From Philosophy of Nature to Subjectivity in the Feminine.* Albany: State University of New York Press, 2002.

Moeller, Hans-Georg. *The Radical Luhmann.* New York: Columbia University Press, 2012.

Monahan, Seth. "Action and Agency Revisited," *Journal of Music Theory* 57, no. 2 (2013): 321–71.

Montagu, Lady Mary Wortley. *Turkish Embassy Letters.* Edited by Malcolm Jack. Athens: University of Georgia Press, 1993.

Morris, Mitchell. "Ecotopian Sounds; or, The Music of John Luther Adams and Strong Environmentalism." In *Crosscurrents and Counterpoints: Offerings in Honor of Bengt Hambraeus at 70,* edited by Per F. Broman, Nora A. Engebretsen, and Bo Alphonce, 129–41. Göteborg, Swed.: Göteborgs Universitet, 1998.

Morton, Timothy. *The Ecological Thought.* Cambridge, MA: Harvard University Press, 2010.

———. *Hyperobjects: Philosophy and Ecology after the End of the World.* Minneapolis: University of Minneapolis Press, 2013.

———. "The Mesh." In LeMenager, Shewry, and Hiltner, *Environmental Criticism for the Twenty-First Century,* 19–30.

———. *Realist Magic: Objects, Ontology, Causality.* Ann Arbor, MI: Open Humanities Press, 2013.

Motraye, Aubry De La. *Travels through Europe, Asia, and into Part of Africa; with Proper Cutts and Maps.* London, 1723.

Motte-Haber, Helga de la. *Musik und Natur: Naturanschauung und musikalische Poetik.* Laaber, Ger.: Laaber-Verlag, 2000.

Müller, Harro. "Luhmann's Systems Theory as a Theory of Modernity." *New German Critique* 61 (1994): 39–54.

Mundy, Rachel. "Birdsong and the Image of Evolution." *Society and Animals* 17 (2009): 206–23.

———. "Evolutionary Categories and Musical Style from Adler to America." *Journal of the American Musicological Society* 67, no. 3 (2014): 737–70.

Nagel, Thomas. *Mind and Cosmos: Why the Materialist Neo-Darwinian Conception of Nature Is Almost Certainly False.* New York: Oxford University Press, 2012.

———. "What Is it Like to Be a Bat?" *Philosophical Review* 83, no. 4 (1974): 435–50.

Narmour, Eugene. *The Analysis and Cognition of Melodic Complexity: The Implication-Realization Model.* Chicago: University of Chicago Press, 1992.

Neff, Severine. "Schoenberg and Goethe: Organicism and Analysis." In *Music Theory and the Exploration of the Past,* edited by Christopher Hatch and David W. Bernstein, 409–33. Chicago: University of Chicago Press, 1993.

Neubauer, John. "Organicism and Music Theory." In *New Paths: Aspects of Music Theory and*

Aesthetics in the Age of Romanticism, edited by Darla Crispin, 11–35. Leuven, Belg.: Leuven University Press, 2009.

Newcomb, Anthony. "Schumann and the Marketplace: From Butterflies to Hausmusik." In *Nineteenth-Century Piano Music*, 2nd ed., edited by R. Larry Todd, 258–315. New York: Routledge, 2004.

Nietzsche, Friedrich. *Beyond Good and Evil*. Translated by Helen Zimmern. In Nietzsche, *Philosophy of Nietzsche*.

——. *The Birth of Tragedy and The Case of Wagner*. Edited and translated by Walter Kaufmann. New York: Vintage Books, 1967.

——. *The Genealogy of Morals*. Translated by Horace B. Samuel. In Nietzsche, *Philosophy of Nietzsche*.

——. *The Philosophy of Nietzsche*. New York: Modern Library, 1937.

——. "Schopenhauer as Educator." In *Unfashionable Observations* (Vol. 2 of *The Complete Works of Friedrich Nietzsche*, edited by Ernst Behler), translated by Richard T. Gray, 169–255. Stanford, CA: Stanford University Press, 1995.

——. *The Will to Power*. Edited by Walter Kaufmann. Translated by Walter Kaufmann and R. J. Hollingdale. New York: Vintage Books, 1968.

Noeske, Nina. "Body and Soul, Content and Form: On Hanslick's Use of the Organism Metaphor." In Grimes, Donovan, and Marx, *Rethinking Hanslick*, 236–58.

Norris, Christopher. "Small Change When We Are to Bodies Gone? Response to Gary Tomlinson." *Opera Quarterly* 29, nos. 3–4 (2014): 203–11.

Novalis [Friedrich von Hardenberg]. *Henry von Ofterdingen*. Translated by Palmer Hilty. Prospect Heights, IL: Waveland Press, 1990.

Parker, Rozsika, and Griselda Pollock. *Old Mistresses: Women, Art and Ideology*. New York: Pantheon Books, 1981.

Pastille, William. "Heinrich Schenker, Anti-Organicist." *19th-Century Music* 8, no. 1 (1984): 29–36.

Patel, Aniruddh D., John R. Iversen, Micah R. Bregman, and Irena Schulz. "Experimental Evidence for Synchronization to a Musical Beat in a Nonhuman Animal." *Current Biology* 19, no. 10 (2009): 827–30.

Peretz, Isabelle. "The Nature of Music from a Biological Perspective." In Peretz et al., "The Nature of Music," 1–32.

Peretz, Isabelle; Ray Jackendoff, and Fred Lerdahl; Sandra E. Trehub and Erin E. Hannon; E. Bigand and B. Poulin-Charronnat; Jamshed J. Bharucha, Megan Curtis, and Kaivon Paroo; and W. Tecumseh Fitch. "The Nature of Music." Special issue, *Cognition* 100, no. 1 (2006).

Phillips-Silver, Jessica. "On the Meaning of Movement in Music, Development and the Brain." *Contemporary Music Review* 28, no. 3 (2009): 293–314.

Phillips-Silver, Jessica, C. Athena Aktipis, and Gregory A. Bryant. "The Ecology of Entrainment: Foundations of Coordinated Rhythmic Movement." *Music Perception* 28, no. 1 (2010): 3–14.

Phillips-Silver, Jessica, and Laurel J. Trainor. "Hearing What the Body Feels: Auditory Encoding of Rhythmic Movement." *Cognition* 105, no. 3 (2007): 533–46.

Piekut, Benjamin. "Chance and Certainty: John Cage's Politics of Nature." *Cultural Critique* 84 (2013): 134–63.

Pijanowski, Bryan C., Luis J. Villanueva-Rivera, Sarah L. Dumyahn, Almo Farina, Bernie L. Krause, Brian M. Napoletano, Stuart H. Gage, and Nadia Pieretti. "Soundscape Ecology: The Science of Sound in the Landscape." *Bioscience* 61, no. 3 (2011): 203–16.

Pollan, Michael. "The Intelligent Plant." *New Yorker*, December 23 and 30, 2013, 92–105.

Prum, Richard O. *The Evolution of Beauty: How Darwin's Forgotten Theory of Mate Choice Shapes the Animal World—and Us*. New York: Doubleday, 2017.

Pryer, Anthony. "Hanslick, Legal Processes, and Scientific Methodologies." In Grimes, Donovan, and Marx, *Rethinking Hanslick*, 52–69.

Purves, Dale. *Music as Biology: The Tones We Like and Why*. Cambridge, MA: Harvard University Press, 2017.

Rampley, Matthew. "Art as a Social System: The Sociological Aesthetics of Niklas Luhmann." *Telos* 148 (2009): 111–40.

Rehding, Alexander. "August Halm's Two Cultures as Nature." In Clark and Rehding, *Music Theory and Natural Order*, 142–60.

Reiman, Erika. *Schumann's Piano Cycles and the Novels of Jean Paul*. Rochester, NY: University of Rochester Press, 2004.

Richards, Robert J. *The Romantic Conception of Life: Science and Philosophy in the Age of Goethe*. Chicago: University of Chicago Press, 2002.

Rigby, Kate. "Earth's Poesy: Romantic Poetics, Natural Philosophy, and Biosemiotics." In Zapf, *Handbook of Ecocriticism and Cultural Ecology*, 45–64.

Riley, Matthew. *Musical Listening in the German Enlightenment: Attention, Wonder and Astonishment*. Aldershot, UK: Ashgate, 2004.

Roosth, Sophia. "Screaming Yeast: Sonocytology, Cytoplasmic Milieus, and Cellular Subjectivities." *Critical Inquiry* 35, no. 2 (2009): 332–50.

Rosen, Charles. *The Classical Style: Haydn, Mozart, Beethoven*. New York: W. W. Norton, 1972.

Rothenberg, David. "Introduction: Does Nature Understand Music?" In Rothenberg and Ulvaeus, *Book of Music and Nature*, 1–10.

———. *Survival of the Beautiful: Art, Science, and Evolution*. New York: Bloomsbury Press, 2011.

Rothenberg, David, and Marta Ulvaeus. *The Book of Music and Nature: An Anthology of Sounds, Words, Thoughts*. Middletown, CT: Wesleyan University Press, 2001.

Rothfarb, Lee A. *Ernst Kurth as Theorist and Analyst*. Philadelphia: University of Pennsylvania Press, 1988.

Rothstein, Edward. "The Americanization of Heinrich Schenker." In *Schenker Studies*, edited by Hedi Siegel, 193–203. Cambridge: Cambridge University Press, 1990.

Rousseau, G. S., ed. *Organic Form: The Life of an Idea*. London: Routledge and Kegan Paul, 1972.

Rousseau, Jean-Jacques, and Johann Gottfried Herder. *On the Origin of Language*. Translated by John H. Moran and Alexander Gode. Chicago: University of Chicago Press, 1966.

Safranski, Rüdiger. *Schopenhauer and the Wild Years of Philosophy*. Translated by Ewald Osers. Cambridge, MA: Harvard University Press, 1990.

Sagan, Dorion. *Cosmic Apprentice: Dispatches from the Edges of Science*. Minneapolis: University of Minnesota Press, 2013.

Scarry, Elaine. "Afterword: An Interview with Elaine Scarry." In LeMenager, Shewry, and Hiltner, *Environmental Criticism for the Twenty-First Century*, 261–74.

Schäfer, Thomas, and Peter Sedlmeier. "Does the Body Move the Soul? The Impact of Arousal on Music Preference." *Music Perception* 29, no. 1 (2011): 37–50.

Schaper, Eva. "Free and Dependent Beauty." In *Kant's Critique of the Power of Judgment: Critical Essays*, edited by Paul Guyer, 101–19. Lanham, MD: Rowman and Littlefield, 2003.

Schenker, Heinrich. *Free Composition*. Vol. 3 of *New Musical Theories and Fantasies*. Translated by Ernst Oster. New York: Longman, 1979.

———. *Neue Musikalische Theorien und Phantasien*. Vol. 3, *Der freie Satz*. Vienna: Universal Edition, 1935.

Scherzinger, Martin. "Anton Webern and the Concept of Symmetrical Inversion: A Reconsideration on the Terrain of Gender." *repercussions* 6, no. 2 (1997): 63–147.

———. "The Return of the Aesthetic: Musical Formalism and Its Place in Political Critique." In *Beyond Structural Listening? Postmodern Modes of Hearing*, edited by Andrew Dell'Antonio, 252–77. Berkeley: University of California Press, 2004.

Schleuning, Peter. *Die Sprache der Natur: Natur in der Musik des 18. Jahrhunderts*. Stuttgart: J. B. Metzler, 1998.

Schmidt, Lothar. "Arabeske: Zu einigen Voraussetzungen und Konsequenzen von Eduard Hanslicks musikalischem Formbegriff." *Archiv für Musikwissenschaft* 46, no. 2 (1989): 91–120.

———. *Organische Form in der Musik: Stationen eines Begriff, 1795–1850*. Kassel, Ger.: Bärenreiter, 1990.

Schopenhauer, Arthur. *Die Welt als Wille und Vorstellung*. 2 vols. Zurich: Haffmans, 1988.

———. *On the Will in Nature*. Edited by David E. Cartwright. Translated by E. F. J. Payne. New York: Berg, 1992.

———. *The World as Will and Representation*. 2 vols. Translated by E. F. J. Payne. Indian Hills, CO: Falcon's Wing Press, 1958.

Schor, Naomi. *Reading in Detail: Aesthetics and the Feminine*. New York: Methuen, 1987.

Schumann, Clara. *Blumenbuch für Robert 1854–1856*. Edited by Gerd Nauhaus and Ingrid Bodsch, with the assistance of Ute Bär and Susanna Kosmale. Bonn: Stroemfeld, 2006.

Schumann, Robert. *Arabeske Op. 18, Blumenstück Op. 19*. Edited by Joachim Draheim. Vienna: Wiener Urtext Edition, 1977.

———. *Early Letters of Robert Schumann*. Translated by May Herbert. London: George Bell and Sons, 1888.

———. *Jugenbriefe von Robert Schumann*. Edited by Clara Schumann. Leipzig: Breitkopf und Härtel, 1885.

———. *Robert Schumann's Briefe: Neue Folge*. Edited by F. Gustav Jansen. Leipzig: Breitkopf und Härtel, 1886.

———. *Robert Schumanns Briefe: Neue Folge*. 2nd ed. Edited by F. Gustav Jansen. Leipzig: Breitkopf und Härtel, 1904.

———. "Vierter und fünfter Quartett-Morgen." *Neue Zeitschrift für Musik* 9, no. 13 (August 14, 1838), 51–52.

Seaton, Beverly. *The Language of Flowers: A History*. Charlottesville: University Press of Virginia, 1995.

Sedlmeier, Peter, Oliver Weigelt, and Eva Walther. "Music Is in the Muscle: How Embodied Cognition May Influence Music Preferences." *Music Perception* 28, no. 3 (2011): 297–306.

Shannon, Claude. "A Mathematical Theory of Communication." *Bell System Technical Journal* 27, no. 3 (1948): 379–423, and no. 4 (1948): 623–56.

Small, Christopher. *Musicking: The Meanings of Performing and Listening*. Hanover, NH: University Press of New England, 1998.

Solie, Ruth. "The Living Work: Organicism and Musical Analysis." *19th-Century Music* 4, no. 2 (1980): 147–56.

———. *Music in Other Words: Victorian Conversations*. Berkeley: University of California Press, 2004.

Sontag, Susan. *Against Interpretation and Other Essays*. New York: Picador, 2013.

Soper, Kate. *What Is Nature? Culture, Politics, and the Non-Human*. Oxford: Blackwell, 1995.

Strauß, Dietmar. *Eduard Hanslick: Vom Musikalisch-Schönen*. Part 1. Mainz: Schott, 1990.

Symanski, Johann Daniel. *Selam, oder die Sprache der Blumen.* 2nd ed. Vienna: Michael Lechner, 1832.

Tadday, Ulrich, ed. *Schumann Handbuch.* Stuttgart: Metzler; Kassel, Ger.: Bärenreiter, 2006.

Tarasti, Eero. "Metaphors of Nature and Organicism in the Epistemology of Music: A 'Biosemiotic' Introduction to the Analysis of Jean Sibelius's Symphonic Thought." In *Musical Semiotics Revisited,* edited by Eero Tarasti, 3–25. Helsinki: International Semiotics Institute, 2003.

Taruskin, Richard. *The Danger of Music and Other Anti-Utopian Essays.* Berkeley: University of California Press, 2009.

———. "Et in Arcadia Ego." In *Danger of Music,* 1–20.

———. "No Ear for Music: The Scary Purity of John Cage." In *Danger of Music,* 261–79.

———. *Oxford History of Western Music.* Vol. 2, *The Seventeenth and Eighteenth Centuries.* New York: Oxford University Press, 2005.

———. *Text and Act: Essays on Music and Performance.* New York: Oxford University Press, 1995.

Thaler, Lotte. *Organische Form in der Musiktheorie des 19. und beginnenden 20. Jahrhunderts.* Munich: Musikverlag Emil Katzbichler, 1984.

Thom, René. "From the Icon to the Symbol." In *Semiotics: An Introductory Anthology,* edited by Robert E. Innis, 272–91. Bloomington: Indiana University Press, 1985.

Thompson, Evan. *Mind in Life: Biology, Phenomenology, and the Sciences of Mind.* Cambridge, MA: Harvard University Press, 2007.

Tischer, Matthias. "Zitat—Musik über Musik—Intertextualität: Wege zu Bakhtin." *Musik und Ästhetik* 13, no. 49 (2009): 55–71.

Titon, Jeff Todd. "The Nature of Ecomusicology." *Música e Cultura* 8, no. 1 (2013): 8–18.

Toiviainen, Petri, and Peter E. Keller. "Spatiotemporal Music Cognition." Special issue, *Music Perception* 28, no. 1 (2010): 1–2.

Tomkins, Silvan. *Affect, Imagery, Consciousness.* 4 vols. New York: Springer, 1962–92.

Tomlinson, Gary. *A Million Years of Music: The Emergence of Human Modernity.* New York: Zone Books, 2015.

———. "Parahuman Wagnerism." *Opera Quarterly* 29, nos. 3–4 (2013): 186–202.

———. "Sign, Affect, and Musicking before the Human." *boundary 2* 43, no. 1 (2016): 143–72.

Treitler, Leo. "The Historiography of Music: Issues of Past and Present." In Cook and Everist, *Rethinking Music,* 356–77.

Truax, Barry. "Soundscape, Acoustic Communication and Environmental Sound Composition." *Contemporary Music Review* 15, nos. 1–2 (1996): 49–65.

Tunbridge, Laura. *Schumann's Late Style.* Cambridge: Cambridge University Press, 2007.

Uexküll, Jakob von. *A Foray into the Worlds of Animals and Humans, with a Theory of Meaning.* Translated by Joseph D. O'Neil. Minneapolis: University of Minnesota Press, 2010.

Varela, Francisco J., Evan Thompson, and Eleanor Rosch. *The Embodied Mind: Cognitive Science and Human Experience.* Cambridge, MA: MIT Press, 1991.

Vila, Anne C. *Enlightenment and Pathology: Sensibility in the Literature and Medicine of Eighteenth-Century France.* Baltimore: Johns Hopkins University Press, 1998.

Waal, Frans de. *Are We Smart Enough to Know How Smart Animals Are?* New York: W. W. Norton, 2016.

———. *Our Inner Ape.* New York: Riverhead Books, 2005.

Wagner, Richard. *Gesammelte Schriften und Dichtungen.* 2nd ed. Vol. 4. Leipzig: E. W. Fritzsch, 1888.

———. *Opera and Drama*. Translated by William Ashton Ellis. Lincoln: University of Nebraska Press, 1995.
Wallin, Nils L. *Biomusicology: Neurophysiological, Neuropsychological, and Evolutionary Perspectives on the Origins and Purposes of Music*. Stuyvesant, NY: Pendragon Press, 1991.
Wallin, Nils L., Björn Merker, and Steven Brown, eds. *The Origins of Music*. Cambridge, MA: MIT Press, 2000.
Watkins, Holly. *Metaphors of Depth in German Musical Thought: From E. T. A. Hoffmann to Arnold Schoenberg*. New York: Cambridge University Press, 2011.
Watkins, Holly, and Melina Esse. "Down with Disembodiment, or, Musicology and the Material Turn." *Women and Music* 19 (2015): 160–68.
———. "Musical Ecologies of Place and Placelessness." *Journal of the American Musicological Society* 64, no. 2 (2011): 404–8.
Weber, Andreas. "Cognition as Expression: On the Autopoietic Foundations of an Aesthetic Theory of Nature." *Sign Systems Studies* 29, no. 1 (2001): 153–67.
Webster, James. "The Eighteenth Century as a Music-Historical Period?" *Eighteenth-Century Music*, vol. 1, no. 1 (2003): 47–60.
———. *Haydn's "Farewell" Symphony and the Idea of Classical Style: Through Composition and Cyclic Integration in His Instrumental Music*. Cambridge: Cambridge University Press, 1991.
Weheliye, Alexander. *Habeas Viscus: Racializing Assemblages, Biopolitics, and Black Feminist Theories of the Human*. Durham, NC: Duke University Press, 2014.
Westerkamp, Hildegard. "Linking Soundscape Composition and Acoustic Ecology." *Organised Sound* 7, no. 1 (2002): 51–56.
———. "Speaking from Inside the Soundscape." In Rothenberg and Ulvaeus, *Book of Music and Nature*, 143–52.
Westling, Louise. "Literature, the Environment, and the Question of the Posthuman." In *Nature in Literary and Cultural Studies: Transatlantic Conversations on Ecocriticism*, edited by Catrin Gersdorf and Sylvia Mayer, 25–47. Amsterdam: Rodolpi, 2006.
Wheeler, Wendy. "The Lightest Burden: The Aesthetic Abductions of Biosemiotics." In Zapf, *Handbook of Ecocriticism and Cultural Ecology*, 19–44.
Wheelock, Gretchen. *Haydn's Ingenious Jesting with Art*. New York: Schirmer Books, 1992.
White, Hayden. *Metahistory: The Historical Imagination in Nineteenth-Century Europe*. Baltimore: Johns Hopkins University Press, 1973.
Whittall, Arnold. "Autonomy/Heteronomy: The Contexts of Musicology." In Cook and Everist, *Rethinking Music*, 73–101.
Williams, Raymond. *Keywords: A Vocabulary of Culture and Society*. Rev. ed. New York: Oxford University Press, 1983.
Wilson, Edward O. *Biophilia*. Cambridge, MA: Harvard University Press, 1984.
Wolfe, Carey. *What Is Posthumanism?* Minneapolis: University of Minnesota Press, 2010.
Zangwill, Nick. *Music and Aesthetic Reality: Formalism and the Limits of Description*. New York: Routledge, 2015.
Zapf, Hubert. *Handbook of Ecocriticism and Cultural Ecology*. Berlin: Walter de Gruyter, 2016.
Zatorre, Robert J., and Isabelle Peretz, eds. *The Biological Foundations of Music*. New York: New York Academy of Science, 2001.
Zuckerkandl, Victor. *Sound and Symbol: Music and the External World*. Translated by Willard R. Trask. Princeton, NJ: Princeton University Press, 1956.

Index

Abbate, Carolyn, 42, 138
Adams, John Luther, 14, 132–39, 143–45, 147–48, 151, 154
Adorno, Theodor: *Aesthetic Theory*, 122; *Dialectic of Enlightenment*, 112, 114–15; and "new musicology," 42; and organicism, 11, 15, 21, 25, 27–28, 33–34, 38
affect: and animal song, 127, 129; Hanslick's suspicion of, 120–21; Herder on, 117–18; musical analogues of, 27; and musical response, 2, 4, 40, 77–79
Alpers, Svetlana, 90
animals: animal communication, 4, 14, 125–29, 135–36, 139; animal studies, 2; in Deleuze and Guattari's philosophy, 124–25; in Goethe's "Theory of Tone," 122; and the grotesque, 60; and music's effects, 13, 114–18, 120; and organicism, 19, 22–23; in Schopenhauer's philosophy, 12, 67–69, 71–72, 75, 78–79, 81–82; in soundscape composition, 145
Anthropocene, 84
anthropocentrism, 4, 11, 39, 68
arabesque, 12, 45, 47–48, 54–56, 112
Aristotle, 37, 81
assemblage theory, 30, 39
Atwell, John E., 70
Atzert, Stephan, 76
autonomy, 18, 23–24, 27–33, 42–44, 47, 63
autopoiesis, 28, 30, 63

Bach, Johann Sebastian, 105, 107
Bateson, Gregory, 2, 138–39
Baumgarten, Alexander, 115
beauty: Kant's theory of, 46–48, 54; musical, 44–47, 49–50, 63–65; natural, 11–12, 48, 81–82
Becker, Judith, 123

Beethoven, Ludwig van: and the cuckoo, 125; modernity of, 31; and organicism, 28–29; *Prometheus* overture, 29, 50, 52; and Schumann, 4; Symphony no. 5, 11, 23
Bennett, Jane, 68
biodiversity, 6–7, 40
biomusicology, 3, 122
biophilia, 130–31
biosemiotics. *See under* semiotics
birdsong, 54, 125–32, 135, 141–42, 152–54
Blumensprache. See flowers: language of
Böhme, Gernot, 83, 170n41, 177n31
Bonds, Mark Evan, 27–28
Bregman, Albert S., 26
Burnham, Scott, 28
Busby, Thomas, 130

Cage, John, 5, 14, 132–38, 144
Chadwick, Whitney, 89
Chua, Daniel, 32
constructivism, 7
Cox, Arnie, 53, 123
Cumming, Naomi, 14, 140, 143
Currie, James, 43

Dahlhaus, Carl, 38
Darwin, Charles, 68, 72, 114, 125
Daverio, John, 39, 85–86, 88, 105, 108
David, Jacques-Louis, 91
Deacon, Terrence, 2, 7, 25, 29, 44, 142
DeLanda, Manuel, 39
Deleuze, Gilles, 2, 39, 58, 124–25, 129, 135, 148
de Man, Paul, 102
DeNora, Tia, 123
Derrida, Jacques, 13, 113–14, 124, 128
Diderot, Denis, 91

Dietzsch, Barbara Regina, 111
Drever, John Levack, 145–46
Dreyfus, Lawrence, 38
Dunn, David, 139, 145

ecomusicology, 3–4
Eichendorff, Joseph von, 152
embodied cognition, 2, 122
embodiment: and communication, 129; in Hanslick's writings, 65; and musical experience, 2, 13–14, 122–26; in Schopenhauer's philosophy, 12, 70–75, 77–80; in Sulzer's writings, 115–17
emergence, 1–2, 25–28, 44, 52–53, 63, 74
ethology, 3, 7, 14
evolution, 4, 72, 79, 83, 128–29
Ewaldt, Karl Wilhelm, 99

Ferrara, Lawrence, 68
flowers: in *Dichterliebe*, 92–94, 101; and formalism, 11; in Hanslick's aesthetics, 44–45, 47, 49; in Hoffmann's writings, 75, 95–96; in Kant's aesthetics, 44, 48; language of, 13, 87, 97–101, 109, 111; in the nineteenth century, 12–13, 87–88; paintings of, 87, 89–92; in Romantic literature, 13, 94–97, 102; and Schumann's *Arabeske*, 60–61; and Schumann's *Blumenstück*, 87–88, 101–2, 111; in Schumann's writings, 88–89; and transience, 65, 111
formalism: and animal song, 126–27, 129; discourse of, 4, 41–46, 122; in Hanslick's writings, 11–12, 45–50, 120–21; and transience, 63–65
Freud, Sigmund, 68

Gagliano, Monica, 66–67, 69
Geissmann, Thomas, 126
gender: and expression, 104, 111; and flowers, 12–13, 87, 89–92, 95–96, 100–101; and masculinity, 18, 104, 107; in the natural world, 6; and politics, 43
Ghosh, Amitav, 80, 83
Goehr, Lydia, 68, 135
Goethe, Johann Wolfgang von: *Märchen*, 96; "Die Metamorphose der Pflanzen," 100; *The Metamorphosis of Plants*, 60; and organicism, 11, 22, 25; on rhythm, 44; and the *salām*, 97; "Theory of Tone," 121–22
Great Chain of Being, 68
Grosz, Elizabeth, 2, 44, 124–25, 128, 148
Guattari, Félix, 2, 58, 124–25, 129, 135

Habermas, Jürgen, 30–31
Hammer, Joseph, 97–98, 111
Hannan, Barbara, 72, 79–80
Hanslick, Eduard: on analysis and criticism, 36, 41–45, 49–50, 52, 54; and the arabesque, 55–56; on beauty, 11, 45–47, 64–65; on listening bodies, 13–14, 114, 118–21, 123–24, 126–27; organicism of, 16, 24, 28–29
Haraway, Donna, 2, 22, 113
Hartshorne, Charles, 128
Hatab, Lawrence J., 10
Hawkins, John, 129
Haydn, Joseph, 31–32, 38
Hayles, N. Katherine, 5
Head, Matthew, 130
Hegel, Georg Wilhelm Friedrich, 22–23, 46, 92, 115, 128
Heine, Heinrich, 13, 92, 94–95, 97, 101
Herder, Johann Gottfried, 13, 48, 117–18, 125–26, 135
Higgins, Kathleen Marie, 1, 129
Hirschbach, Hermann, 94
Hoffmann, Ernst Theodor Amadeus: and flowers, 13, 66, 95–96; and the infinite, 48; musical aesthetics, 15, 35, 75–76; organicism of, 11, 15, 23–24
Hölderlin, Friedrich, 102
Horkheimer, Max, 112, 114
human exceptionalism, 3, 40, 69, 124, 128
humanism, 3, 5–6, 8, 43–44, 125

information theory, 137, 142

Jean Paul (Richter), 87, 89, 92, 94, 100, 111

Kaminsky, Peter, 103
Kant, Immanuel: aesthetics of, 44–54, 58, 82, 115; and the "beautiful soul," 154; organicism of, 11, 21–23, 25, 28, 38–39
Kantianism, 10, 45
Kawohl, Friedemann, 40
Keller, Peter E., 122
Kerman, Joseph, 42
Kircher, Athanasius, 129
Kittler, Friedrich, 13, 95–96, 100–102, 107, 152
Kohn, Eduardo: on emergence, 27; on form, 44; on "living thought," 153; semiotics of, 2, 14, 134–36, 139, 142–44, 148
Kramer, Lawrence, 42
Krieger, Murray, 40
Kurth, Ernst, 42, 53

Lacan, Jacques, 13, 113, 118, 124, 126, 128
Langer, Susanne, 2, 36, 44, 53, 58
Larson, Steve, 52
Latour, Bruno, 7, 133
Latour, Charlotte de, 99–100
Leibniz, Gottfried Wilhelm, 77
Lemm, Vanessa, 10–11
Leyster, Judith, 89
Lucretius, 129
Luhmann, Niklas, 2, 11, 19–20, 27, 29–33, 37–39, 63

INDEX

Mabey, Richard, 52
Magee, Bryan, 75
Malabou, Catherine, 7
Malo, Charles, 88
Mancuso, Stefano, 20
Marder, Michael: and critical plant studies, 2; on Gagliano, 66–67, 69; on Goethe, 60; on "plant-thinking," 24, 40, 55, 58
Marler, Peter, 126–28, 136
Martinelli, Dario, 137
Marx, Adolf Bernhard, 16, 24
materialism, 5, 12, 68–69
Maturana, Humberto, 28–30
McClary, Susan, 42
Merian, Maria Sibylla, 89
Merker, Björn, 127
Messiaen, Olivier, 125
metabolism, 1, 53, 55–56
metaphor, 36–40, 94
Meyer, Leonard, 27
Michaelis, Christian Friedrich, 11, 25–26, 28
Michelangelo, 89
Montagu, Lady Mary Wortley, 97–98
Montausier, Marquis de, 88
Morton, Timothy, 18
Motraye, Aubry De La, 97
Motte-Haber, Helga de la, 25
Mozart, Wolfgang Amadeus, 31
Miller, J. M., 100
Müller, Harro, 31
music perception, 3, 122–24

Narmour, Eugene, 27
nature: and aesthetics, 11–12, 43–44, 46–48, 50, 112; and humanity, 6–7, 11; and modernism, 132–34; in Nietzsche's philosophy, 9–10; and *physis*, 7–8, 11; and posthumanism, 6; in Romantic literature, 95–96; in Schopenhauer's philosophy, 67–69, 76, 79–84, 111; and self-organization, 50–53; semiotics of, 134–39; and transience, 65
Neubauer, John, 39
Newcomb, Anthony, 85
Nietzsche, Friedrich, 8–11, 76, 115, 124
Novalis (Friedrich von Hardenberg), 94–97, 99

organicism: discourse of, 4, 11, 13, 15–26; and dynamic systems, 32–34; in Goethe's "Theory of Tone," 121–22; in Hanslick's writings, 45, 50; in Langer's writings, 55; and motivic relations, 103, 108–9; and musicology, 35–40, 42; and self-organization, 26–29; and teleology, 52–53; in Schumann's *Arabeske*, 60, 63
Orpheus, 112, 116, 120

Peircean semiotics. *See* semiotics: Peircean
periodicity, 3, 55–56, 58, 112

Phillips-Silver, Jessica, 122
physis, 5, 7–8, 11, 21
Piekut, Benjamin, 133
plants: and arabesque, 12, 55; and beauty, 54, 82; critical plant studies, 2; and organicism, 4, 11, 19–24, 39–40; and Schopenhauer's philosophy, 12, 66–69, 71–72, 74–75; and Schumann's *Arabeske*, 58, 60
Platner, Anton, 100–101
Platonic Ideas, 69, 72–73, 81
Pollan, Michael, 19–20
posthumanism, 5–8
Printz, Wolfgang Caspar, 129

Reiman, Erika, 86, 89
Riley, Matthew, 116
Romanticism: and flowers, 13, 87, 94–102, 109; Hanslick's opposition to, 45; and Schopenhauer's thought, 69, 75–76
Rothenberg, David, 137, 139
Rousseau, Jean-Jacques, 91
Ruysch, Rachel, 89

Safranski, Rüdiger, 68
Schaeffer, Pierre, 14, 137
Schafer, R. Murray, 154
Schäfer, Thomas, 123
Schenker, Heinrich, 16, 38, 42, 53
Scherzinger, Martin, 42–43
Schiller, Friedrich, 100
Schmidt, Lothar, 16
Schoenberg, Arnold, 34
Schopenhauer, Arthur: and nature, 111; philosophy of, 4, 12, 14, 48, 66–84, 96; and posthumanism, 8
Schor, Naomi, 103
Schubert, Franz, 100
Schumann, Robert: *Arabeske*, op. 18, 13, 45, 57–63, 85; *Blumenstück*, op. 19, 12–13, 85–89, 92, 100–111; *Dichterliebe*, op. 48, 92–94, 101; *Humoreske*, op. 20, 85; *Kinderszenen*, op. 15, 61–63; *Kreisleriana*, op. 16, 85, 105; music of, 4; *Myrthen*, op. 25, 99; *Waldszenen*, op. 82, 14, 133–34, 148–54
Seaton, Beverly, 94, 97
Sedlmeier, Peter, 122–23
self-organization: in music, 26–29, 52–53, 55, 63, 65; in organisms, 22, 45, 51; in systems theory, 19–20, 26, 38
self-similarity, 26, 52, 54–56, 58
semiotics: biosemiotics, 3, 14, 56, 134–44; discourse of, 2; of music, 133–35, 139–44; Peircean, 14, 135–36, 140, 142–43; of Schumann's *Waldszenen*, 148–54; of sound, 135–39; of soundscape composition, 144–48
sexuality, 6–7, 43, 125

Shannon, Claude, 137
Sontag, Susan, 138
Soper, Kate, 10
Sulzer, Johann Georg, 13, 114–20, 129
Symanski, Johann Daniel, 100
systems theory: coupling in, 9, 27, 30, 40, 46, 54; discourse of, 2, 9, 11, 19–20, 26–35, 46, 138; recursion in, 30–33; and second-order observation, 32; and third-order observation, 63

Taruskin, Richard, 42, 53, 137
techne, 5, 7–8, 11, 21
teleology, 44–45, 50–53
Thaler, Lotte, 16, 18
Thompson, Evan, 2, 25, 28–29, 33, 51–52
Toiviainen, Petri, 122
Tomlinson, Gary, 14, 128–30, 134, 139–44, 149
Trainor, Laurel J., 122
Treitler, Leo, 39
Truax, Barry, 146
Tunbridge, Laura, 152

Uexküll, Jakob von, 10, 142

Varela, Francisco, 2, 28–30
virtuality, 2, 79, 140, 144–48, 151
vitalism, 53, 157n7
vitality, 1–4, 11, 14, 20, 44–45, 53, 56

Waal, Frans de, 113, 136
Wagner, Richard, 33–34; and symbolic representation, 139, 142; *Tristan und Isolde*, 11, 15–19, 21, 33–37
Wallin, Nils L., 27–28
Walther, Eva, 122
Weber, Andreas, 56, 63
Webster, James, 32, 38
Weigelt, Oliver, 122
Westerkamp, Hildegard, 145–47
White, Hayden, 39
wholeness, 19, 21–26, 28–29, 33, 38
Wieck, Clara, 88, 92, 99
Wilson, E. O., 131
Winderen, Jana, 14, 134, 144–45

zoomusicology, 122
Zuckerkandl, Victor, 53